AF594814

# PAINTING SURF & SEA

# PAINTING SURF & SEA

Harry R. Ballinger

DOVER PUBLICATIONS, INC.
Mineola, New York

*Bibliographical Note*

This Dover edition, first published in 2008, is a republication of the work originally published by Watson-Guptill Publications, New York, in 1957. The original color plates have been gathered together as an eight-page color insert between pp. 42 and 43. These plates also appear in black and white in their original positions in the book.

*Library of Congress Cataloging-in-Publication Data*

Ballinger, Harry Russell, 1892–1993.
Painting surf and sea / Harry R. Ballinger.
p. cm.
Originally published: New York : Watson-Guptill Publications, 1957.
ISBN-13: 978-0-486-46427-5
ISBN-10: 0-486-46427-X
1. Ocean waves in art. 2. Sea in art. 3. Marine painting—Technique. I. Title.

ND1370.B3 2008
751.45'437—dc22

2007043851

Manufactured in the United States of America
Dover Publications, Inc., 31 East 2nd Street, Mineola, N.Y. 11501

*To Kay*
*who has always wanted*
*me to write this book.*

# Contents

# Introduction

I think that almost everyone loves the sea and, while watching the great swells come crashing into shore, has had an irresistible impulse to try to capture some of its beauty and majesty with brush or crayon.

Water by its very nature, in constant motion, is hard to paint. When once you understand the principle on which the sea and surf operates, you will find that it is no longer confusing—it conforms to an orderly pattern and is easy to understand.

Recently, I have seen some fascinating attempts at marine painting done by artists who probably loved the sea but didn't know a thing about it. In some of these pictures, the cresting waves looked like ostrich plumes. In others the waves appeared frozen and as hard as a rock; they couldn't have been dented with a hammer.

I have always thought it a shame that there was so little general information about marine painting available to the average artist who liked the sea and wanted to paint it. It seemed to me that there was a great need for a down-to-earth book that would be a guide for all prospective marine painters—one that would explain clearly just what was happening in the confusing action of surf breaking on the shore and that would help one to see the pattern on which waves and surf operate. I have searched for just such a simply written but authoritative book on marine painting, but could never find one. None seemed to give enough concrete information on the subject or was written simply enough to be easily understood by the beginner or average painter.

After meditating on the problem for a number of years I decided that I would write a book on marine painting myself. One that would explain in simple nontechnical language all the problems involved in painting a seascape and that would be profusely illustrated. Personally I like a book to be so completely illustrated that one can enjoy it without bothering to read the text!

It seems to me that most books written by artists are primarily designed to astonish and impress their fellow painters. It would be a fine thing to write a scholarly tome that would be the last word on sea painting, but I still think that what we need today is a book which is easy to read and to understand and, of course, has plenty of pictures.

So here is a book which contains almost all of the helpful ideas that I have learned by trial and error in nearly thirty years of painting the sea. It also shows step by step how to go about painting a seascape and contains many tips on picturemaking in general.

If this book helps any of you nice people to a better understanding of the sea and how to paint it, I will feel amply repaid for the effort I have made to write and illustrate *Painting Surf and Sea.*

Harry Ballinger
New Hartford, Connecticut
March 1, 1957

# 1: Painting Equipment

For the benefit of those who have had no previous experience with oil painting, I will list the equipment needed for out-of-door painting.

First you will need a sketchbox, either a size 12 x 16 inches or, if you want a more generous one, 16 x 20 inches. I think one which is smaller than 12 x 16 is too cramped for comfort unless one happens to be a midget or a miniature painter. The box can be made either of wood or aluminum. Personally I like a wooden box, but a great many painters prefer the aluminum type because of its light weight. A palette comes with the sketchbox, though some painters may prefer a paper palette with disposable sheets. If you have a wooden one, it is a good idea to rub a little linseed oil on the palette when new, to fill up the porousness of the wood and to give a smoother surface on which to mix your paints. When buying a sketchbox be sure that the dealer does not sell you a fitted box, as they generally contain a weird assortment of earth colors and exotic tints that haven't been in general use since the days of the Pharaohs (Fig. 1).

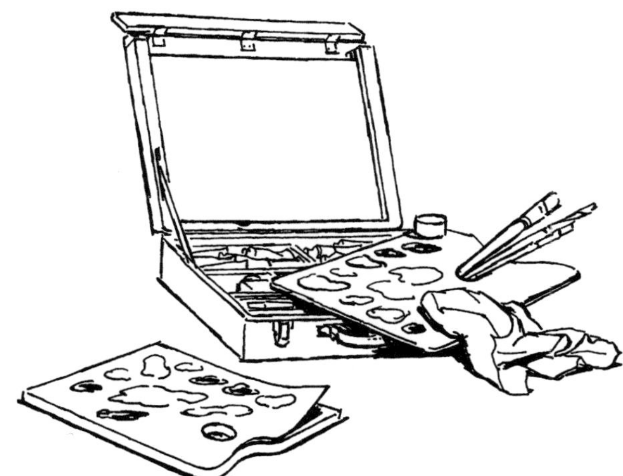

Fig. 1. *Sketch box and disposable palette*

You will need a good solid sketching easel, either of wood or aluminum. The type of easel I prefer is the one that is manufactured by Edith Anderson Miller in Cincinnati, Ohio (formerly made by Oscar Anderson). It has a shelf-like arrangement on which to rest your sketchbox and palette below your canvas. Some of the aluminum easels have a similar arrangement for resting your palette and are equally good. If possible, get an easel with a place to rest your palette as it is extremely inconvenient to have to hold your palette in your left hand. It is better to have your left hand free to hold a paint rag and a few extra brushes (Fig. 2).

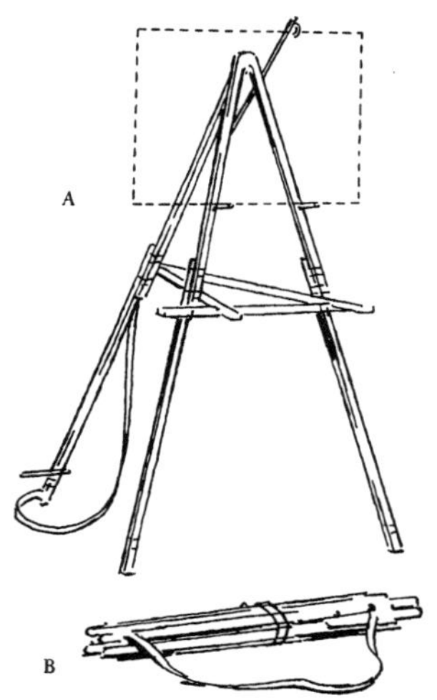


Fig. 2. *Anderson Easel—(A) in use (B) folded up ready for carrying*

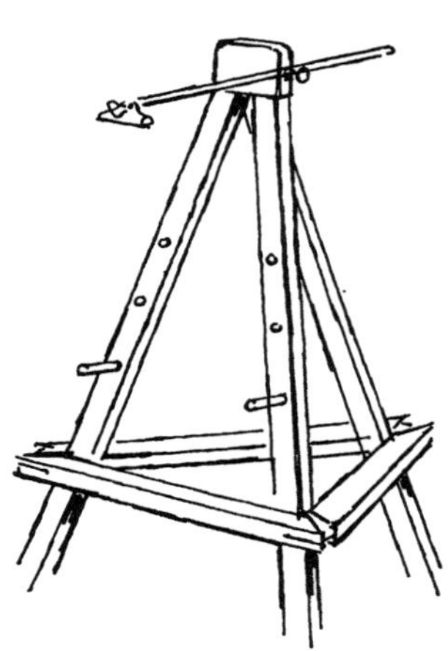

Fig. 3. *Three Stretcher stick nailed together to make triangle*

The Anderson-type easel has a shoulder strap for carrying it, which is a great help when you are climbing over the rocks with a paint box in one hand and a couple of canvases in the other.

If your easel hasn't a place for your palette, you can nail three stretcher sticks together into a triangle that will fit down over the easel and give you a platform for the palette (Fig. 3).

You will need at least six brushes, ranging from about a quarter-inch to an inch in width. I prefer bristle brushes rather than sable, as you are liable to get your picture too slick and polished-up with sable brushes. The brushes should be flat, with a square end, and can be either the type called *brights,* with a short bristle, or the ones with a slightly longer bristle called *flats.* I would not advise brushes with extremely long bristles, as they are difficult to manipulate (Fig. 4).

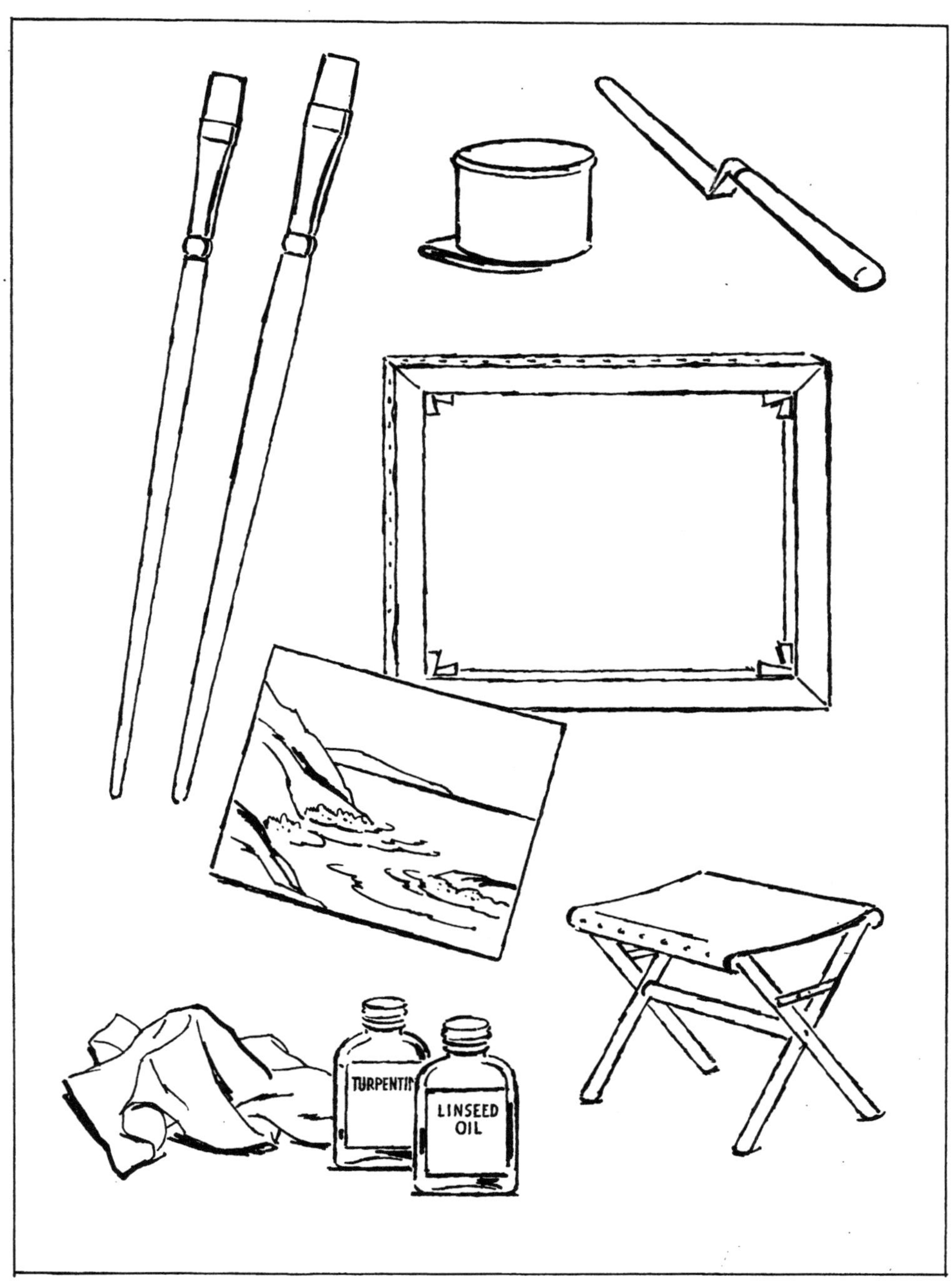


Fig. 4. *All necessary painting equipment*

It is a good idea to get two brushes of approximately the same size, one for light colors and the other for dark. For your smaller sizes, Nos. 2 and 3; for the next size 4 and 5; and for the large size 6 or 7 or possibly 8. For very fine lines you might have a No. 1 or 2 brush, but you will find you can get a fairly fine line by using the edge of the brush and will seldom need a really small brush. It is a good plan to get used to using as large a brush as possible in your picture, as this will make it easier for you to cover the canvas with paint. It will also help you to keep your work broad and simple.

You will need an oil cup for holding your painting medium. It should be about two inches in diameter—a single cup is all you need.

You will also need a palette knife, which should be the trowel-like type with the knife surface a little below the handle. This is somewhat easier to use than the straight type, and you get less paint on your hands in using it.

The list of colors that is used in what has flatteringly been called the *Ballinger palette* is a simple one.

It consists of ultramarine blue and cerulean blue; zinc white; cadmium yellow pale, cadmium orange, cadmium red light, cadmium red deep, and alizarin crimson. There are no earth colors, greens or black.

These are the colors that you really need for most of your painting. In addition you can buy phthalo blue or viridian, but you will find that you rarely use them.

There are a few terms that artists use in describing the painting of a picture which may be a little confusing to the beginner. Artists are constantly talking about *warm* and *cool colors* in their pictures. This means exactly what it says. By warm ones we mean red-yellow-orange; yellowish greens, reddish purples, and browns or grays in which the warm colors predominate. The cool colors are blue, blue-green, bluish purple, and the grays and dark colors in which cool colors predominate.

By *values* we mean the degree of light and dark of any particular part of a picture or color. It could also mean the degree of light or dark of one color compared with another.

*Tone* means about the same as value.

*Key* means the color values of the painting. A high-keyed painting would be a picture with light, bright colors; a low-keyed one would be a picture with dark, somber colors.

*Medium* is the liquid which you use to mix with your paint so that you can apply it to the canvas more readily. I use a mixture of linseed oil and turpentine—half-and-half. The oil should be a purified linseed oil, obtainable from an artists' materials store, but any clear, good gum turpentine sold by paint stores is adequate.

You will also need retouching varnish. This is a quick-drying varnish to bring a gloss to the parts of the picture that look dull and lifeless. It can be sprayed on by a fixative blower if the painting is wet, or applied with a clean soft brush if the paint is dry. You will need to use the retouching varnish only when the picture has dried out and you wish to continue painting on it.

The last item is something to paint on. Canvas mounted on a cardboard panel is good, as it is easy to transport. For any picture larger than 20 x 24 inches, a stretched canvas is better. Panels the size of your sketchbox are good because you can carry them in the lid of your box, which has slots made for that purpose.

If you are painting on stretched canvas, it is a good idea to carry two canvases of the same size with you so that you can put one in back of the other on your easel. This will prevent the sun from shining through the canvas on which you are painting.

For those who prefer to paint sitting down, any strong light folding stool will be satisfactory.

# 2: Analysis of Wave Action

The most important part of a marine painting is the water itself. Until you know just what changes a wave goes through when it breaks on shore, you can't even think about trying to paint a seascape.

In order to paint a surf scene convincingly, it is necessary to understand what is happening before your eyes. The surf moves so rapidly that it is hard to see the action unless you know exactly what to look for. Therefore, I will explain each step in the action of a wave from the time it approaches the shore until it spends its force against the shore and pours back out to sea as floating foam.

Waves move into shore in parallel lines from the open sea. They are evenly spaced, and crest as they get into shallow water at about the same spot each time (Fig. 5).

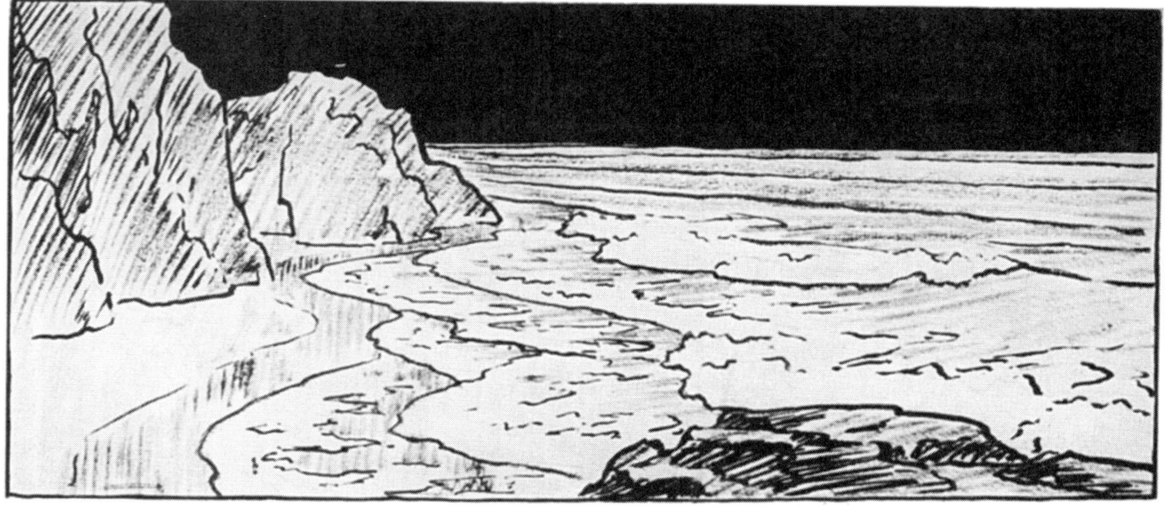

Fig. 5. *Diagram of direction of waves moving into shore*

A cross section of a wave approaching the shore is a triangle with the steepest side toward shore. As a wave reaches shallow water, the front plane

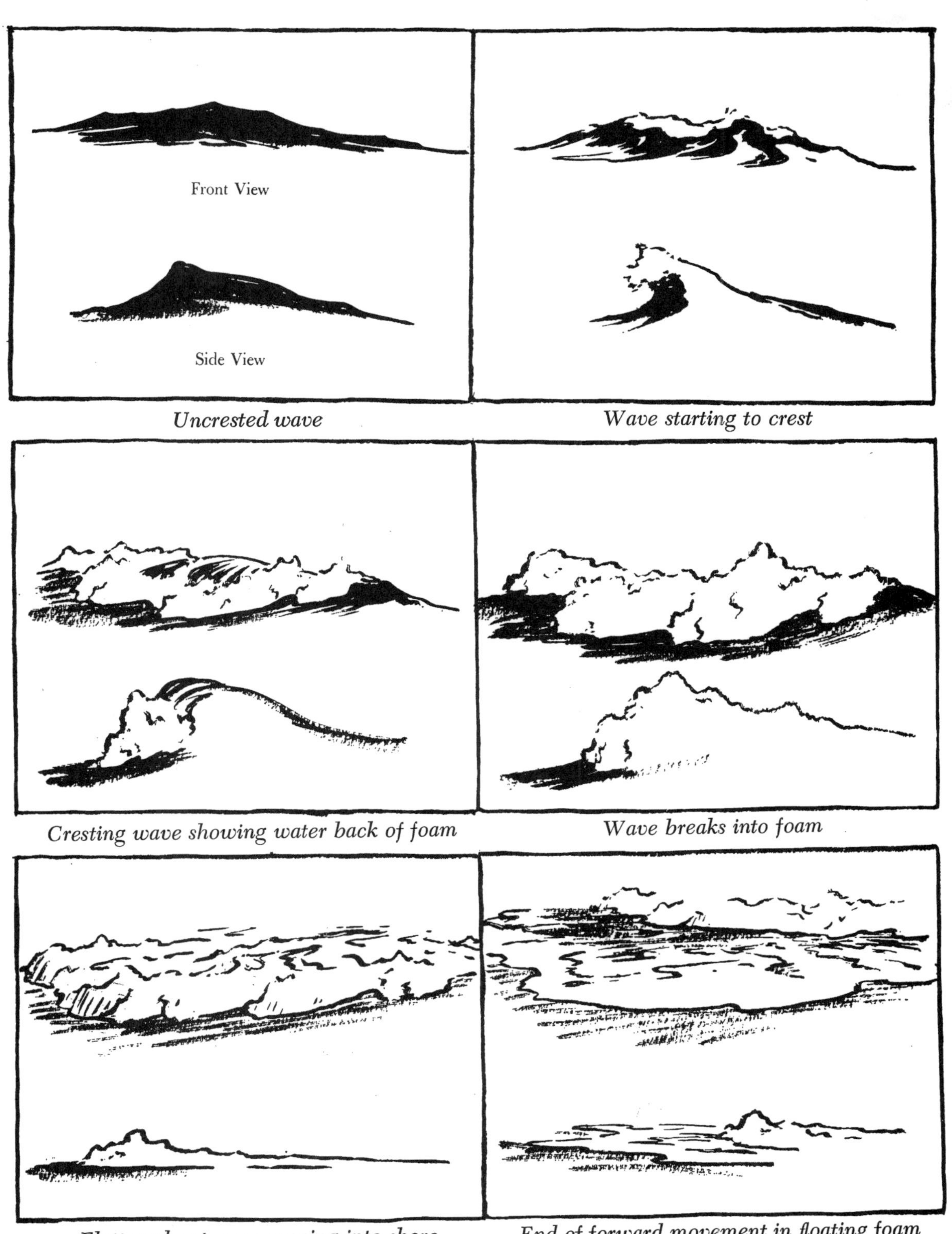


*Uncrested wave* — *Wave starting to crest*

*Cresting wave showing water back of foam* — *Wave breaks into foam*

*Flattened-out wave moving into shore* — *End of forward movement in floating foam*

Fig. 6. *Action of a wave going into shore*

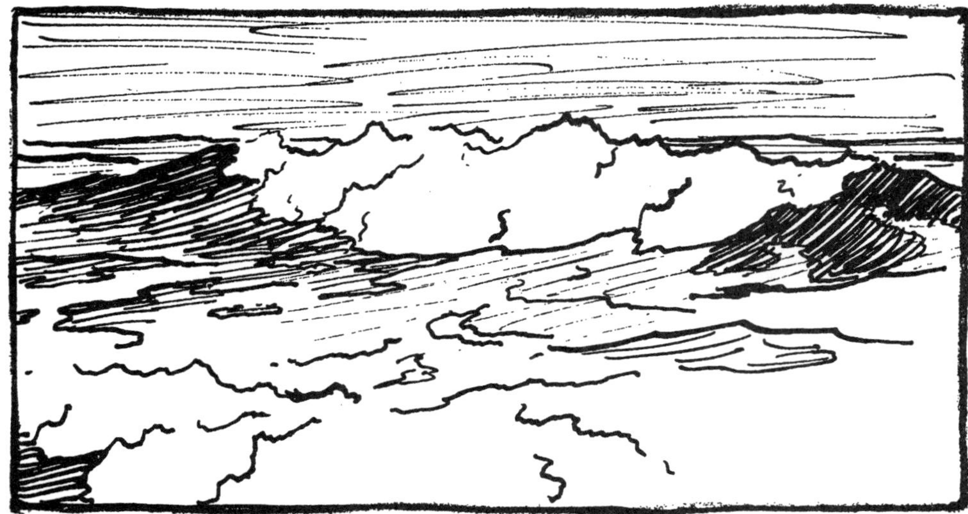

Fig. 7. *Diagram of cresting wave (front view)*

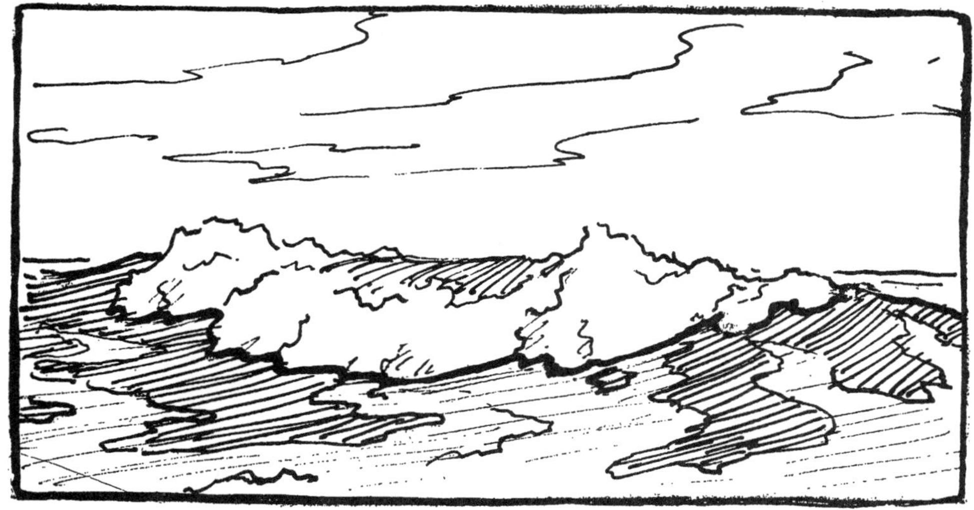

Fig. 8. *Diagram of bottom of a cresting wave*

toward shore gets steeper and steeper until finally the top of the wave collapses forward, rolls over and down, hitting the front slope of the wave near the base with a mighty rebound. The weight of the water behind it forces the foam on toward shore in a series of bouncing masses.

When the top of the wave rolls forward and over, and breaks into foam, there is briefly visible a sheet of water back of the foam that is almost immediately blown apart by the imprisoned air under it. Because of the foam and air bubbles underneath, this sheet of water is often a lovely, pale jade-green in color.

When the sheet of water at the top of the cresting wave disappears, there is left a large mass of foam that comes cascading down the front of the wave in terraces. As the wave moves on toward shore, it gradually flattens out and leaves a twisted mass of foam trailing out behind it. The wave gets flatter as it moves toward shore, though there is still weight and thickness to the mass of foam as long as the forward movement continues. When the wave has traveled up and over the rocks as far as its momentum will carry it, the foam then levels off and pours back down to the level of the sea, directly in front of the rocks.

Sometimes the mass of water pouring off the rocks is so heavy that it gives the effect of a flattened-out wave heading back out to sea. When this backwash hits an advancing wave, you sometimes get quite an impressive splash. Personally, I think that it is confusing to play up the backwash too much in a picture. It rather detracts from the power that you get in your painting when you have all the waves moving in toward shore in the same general direction.

On a flat, sandy beach there is very little backwash noticeable. When the wave travels up the beach and reaches the end of its forward movement, it turns into floating foam and gradually moves out to sea again.

The drawings on page 15 show the action of a wave from the time it crests and moves into shore until it finally pours back out to sea again (Fig. 6).

A wave crests at its highest point when it gets into shallow water (Fig. 7). The foam spreads out toward each end of the wave. This action is especially easy to see when you watch surf on a sandy beach. There the waves, because of the uniform depth of the water, generally show a wide stretch of foam when cresting and the waves appear much longer from end to end than on a rocky shore with its varying depths of water.

When painting a breaking wave coming toward you, notice that the bottom edge of the foam tumbling down the face of the wave can be suggested by sweeping curves deeper at the middle and tapering out toward the ends of the wave (Fig. 8). A profile view of a breaking wave suggests a series of terraces from the top to the bottom in a convex contour rather than

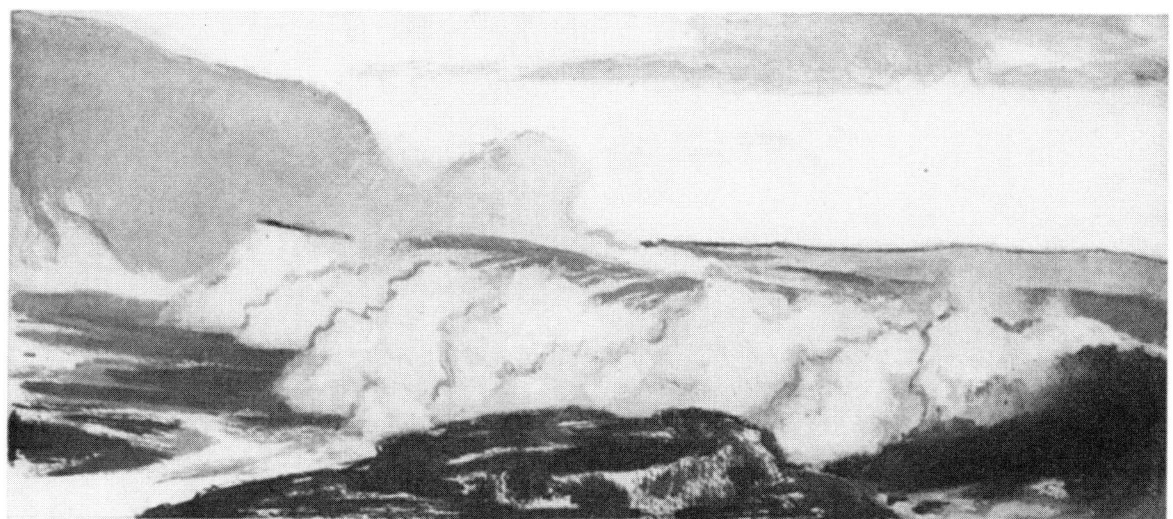

Fig. 9. *Profile view of cresting wave*

the concave shape that some painters imagine a cresting wave assumes (Fig. 9).

A wave doesn't really roll over completely when it crests; just the top portion rolls over and down (Fig. 10). The foam then goes bouncing along toward shore, pushed along by the volume of water behind it and getting flatter and more spread out as it goes along. I always paint these masses of foam with up and down strokes of my brush across the action of the foam itself. The floating foam should be painted with horizontal strokes in the direction in which the foam is moving. This will make it lie flat on the surface of the water (Fig. 11).

If you study the drawings on the opposite page, I think you can easily understand the system on which surf operates. There will be more about this in the chapters on Painting Surf on a Rocky Shore, and Painting Surf on a Sandy Beach.

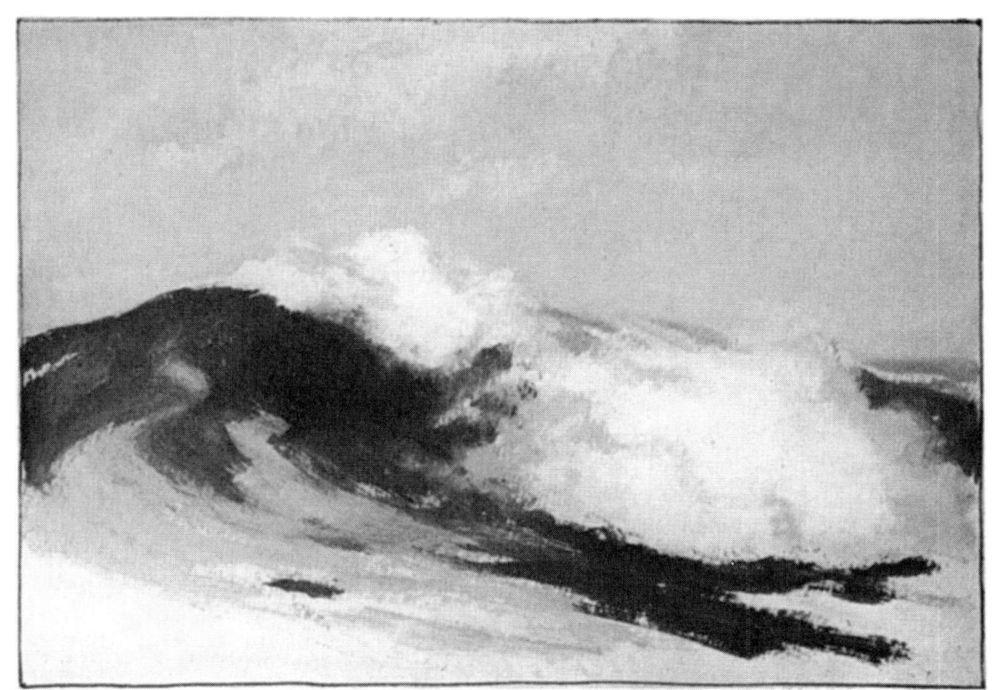

Fig. 10. *Action of the top of a cresting wave*

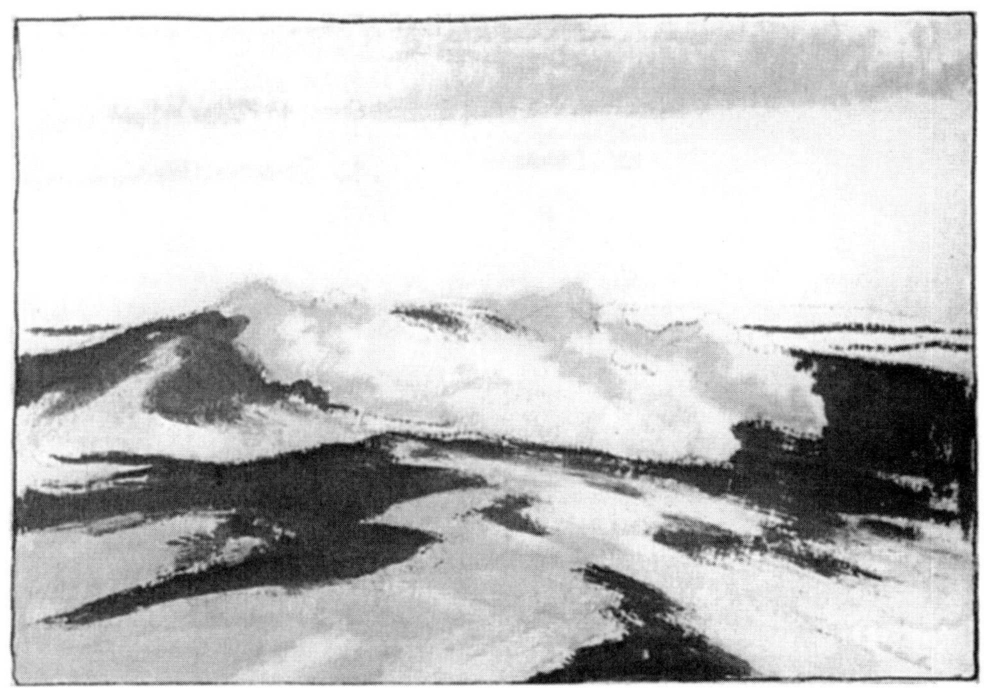

Fig. 11. *Showing method used in painting floating foam*

# 3: Composition

Before starting a marine painting, I would like to explain a few fundamental ideas of composition that apply to all kinds of painting. I think the first in importance is the black-and-white pattern or design of the picture. I always try to think of every scene I paint as a big, simple arrangement in two tones of light and dark (Fig. 12). This is really the framework for your whole picture. I try to see every portion of the picture either as part of the dark or light pattern. I always try to tie my darks together by having one blend into another, to make a large irregular shape of dark rather than a number of isolated dark spots (Fig. 13A & B). The dark pattern, of course, makes the light one.

By thinking of your picture as a two-toned pattern of light and dark, you start with a simple, poster-like design (Fig. 14). Then, as you carry on the picture, you can modify some of the darks and lights; but be careful not to lose your original simple design of light and dark (Fig. 15).

In a seascape, it is easy to see a two-toned pattern of light and shade.

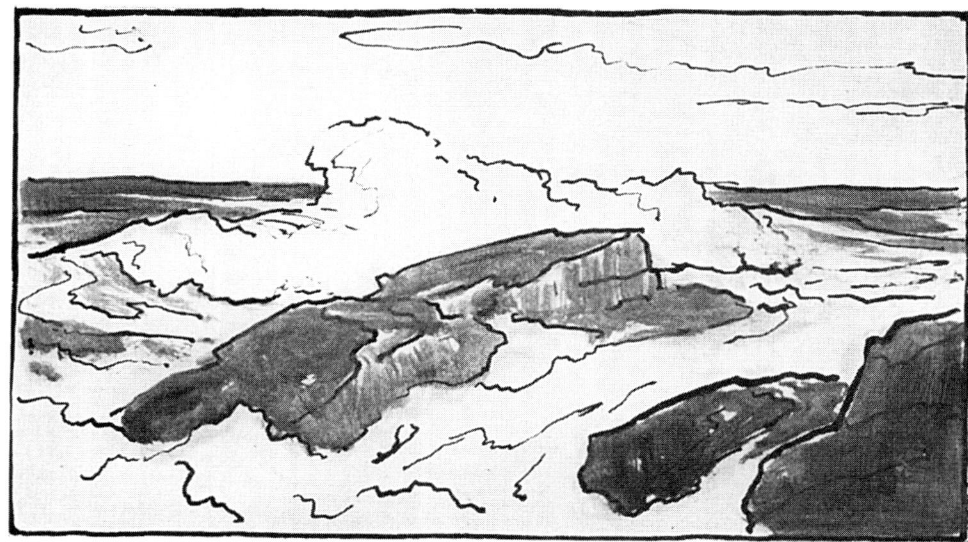

Fig. 12. *Two-toned pattern of composition*

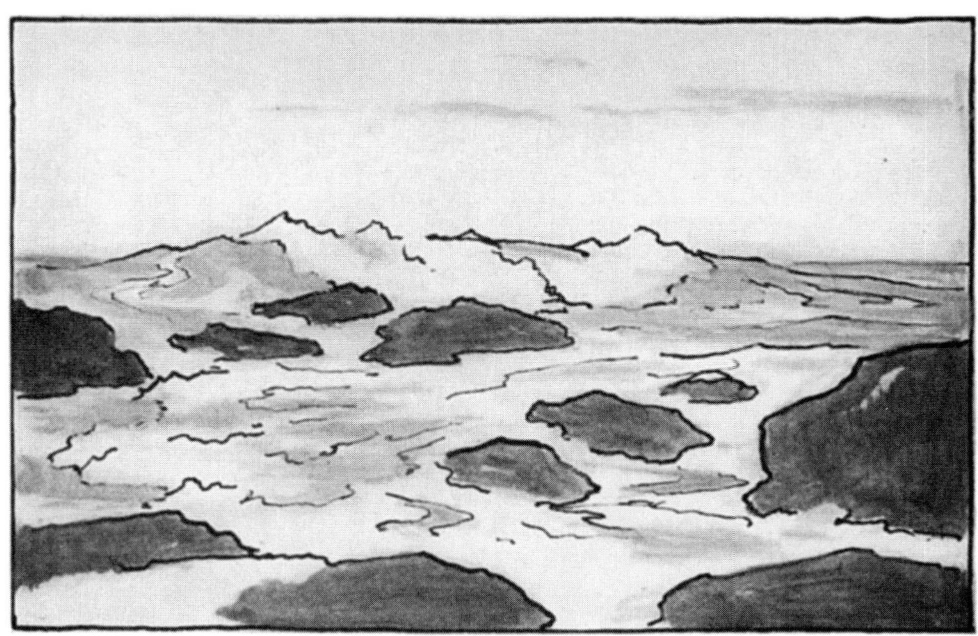

Fig 13A. *Scattered spots of light and dark*

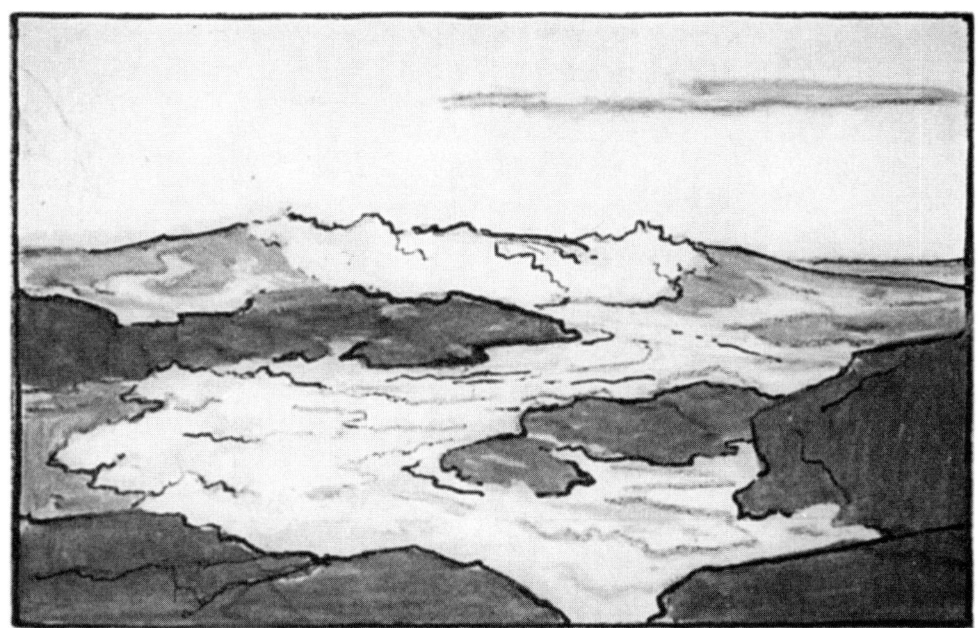

Fig. 13B. *Tying up isolated spots of light and dark*

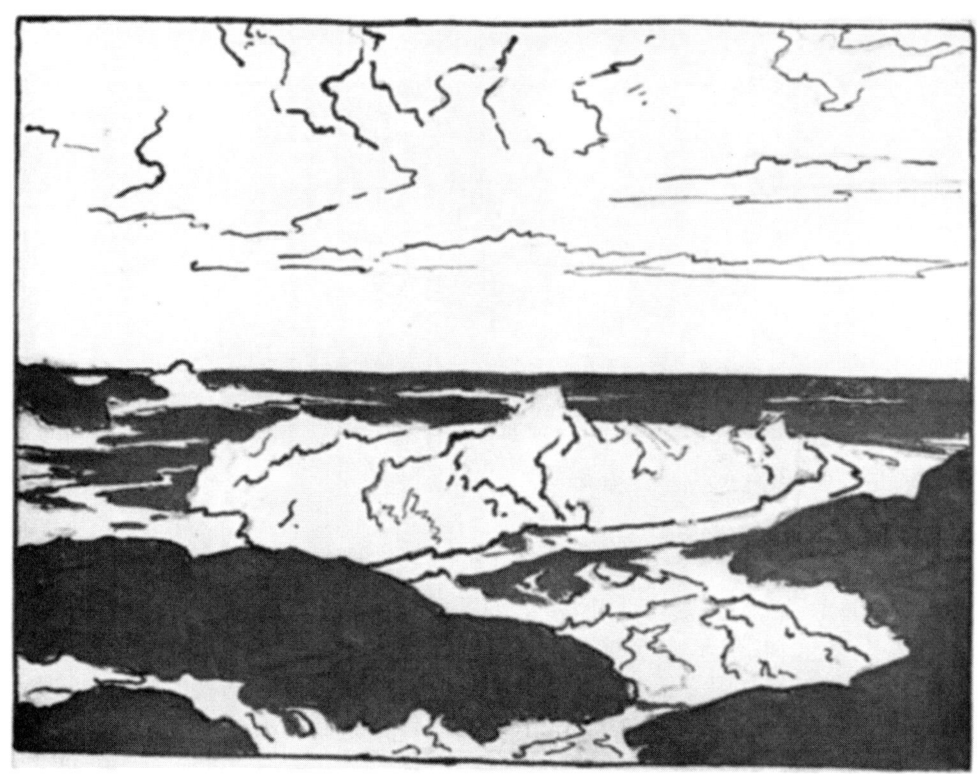

Fig. 14. *Two-toned pattern of picture—first step*

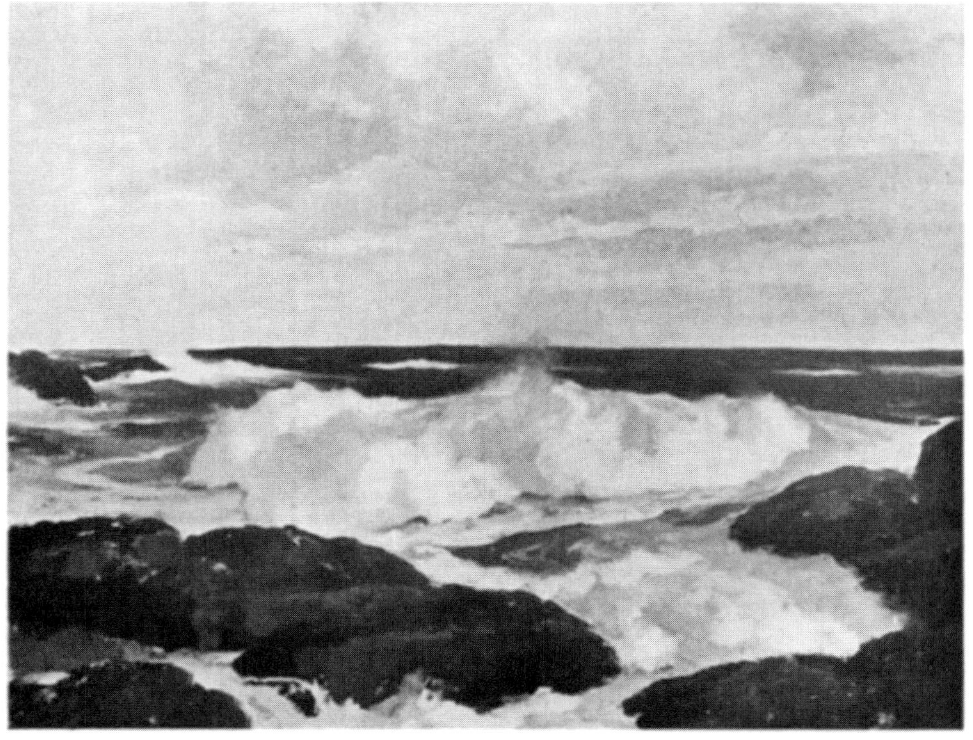

Fig. 15. *Second stage of picture; values worked out*

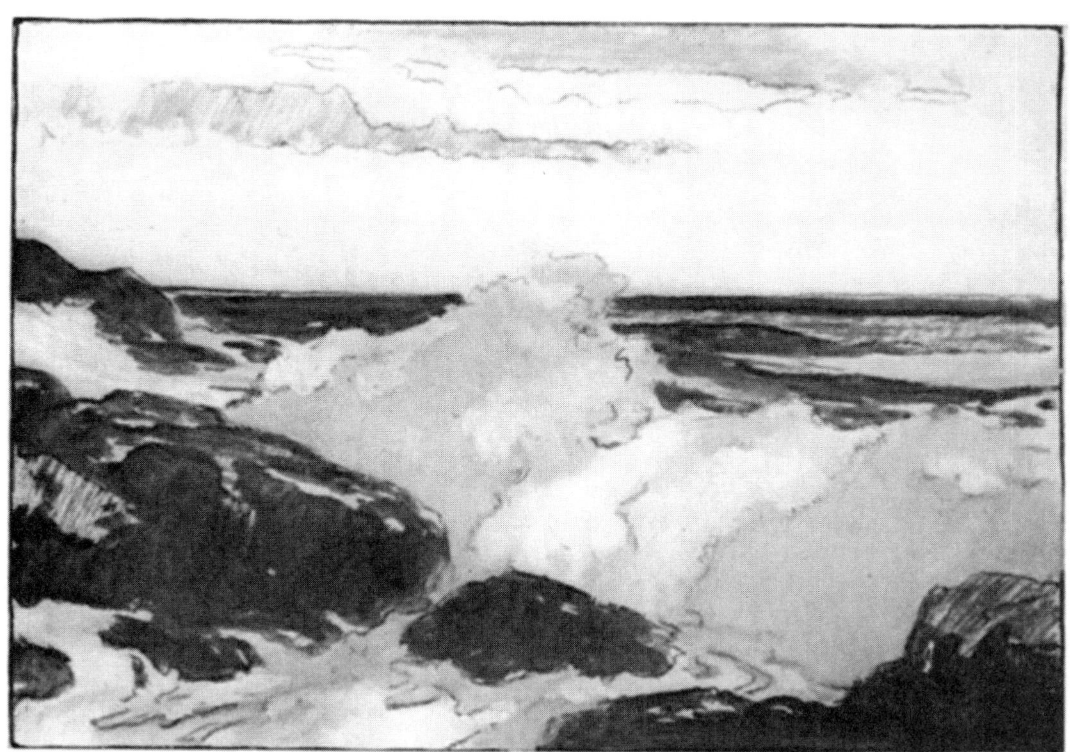
Fig. 16A. *Seascape with diagram of black-and-white pattern of composition*

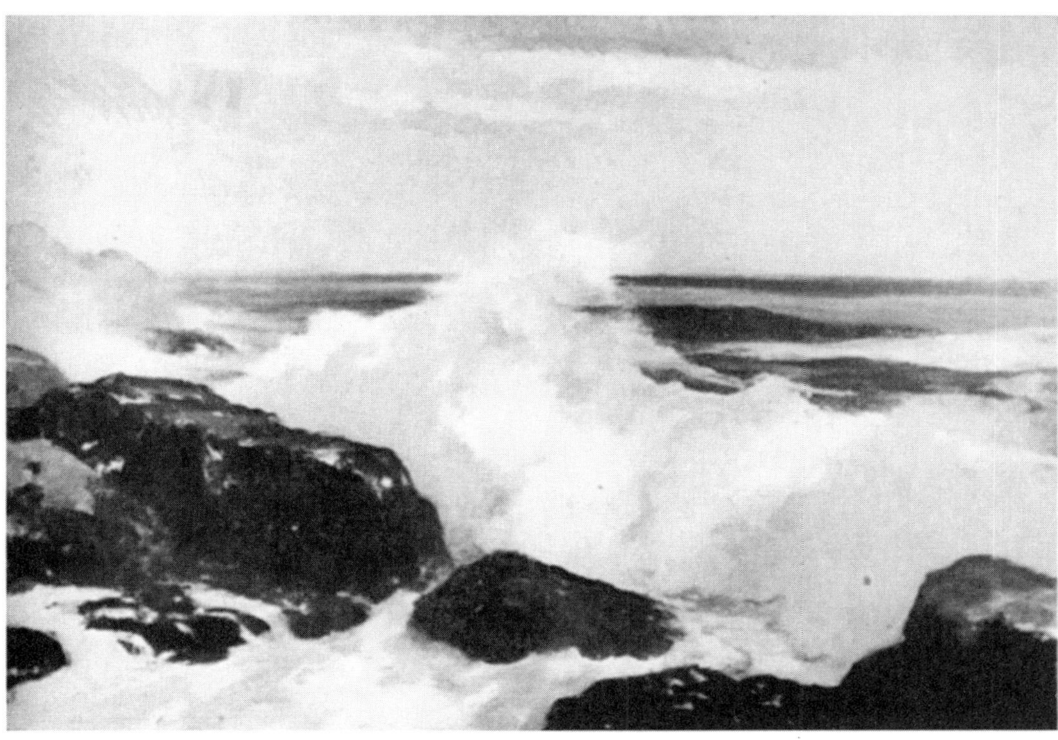
Fig. 16B. *Completed picture*

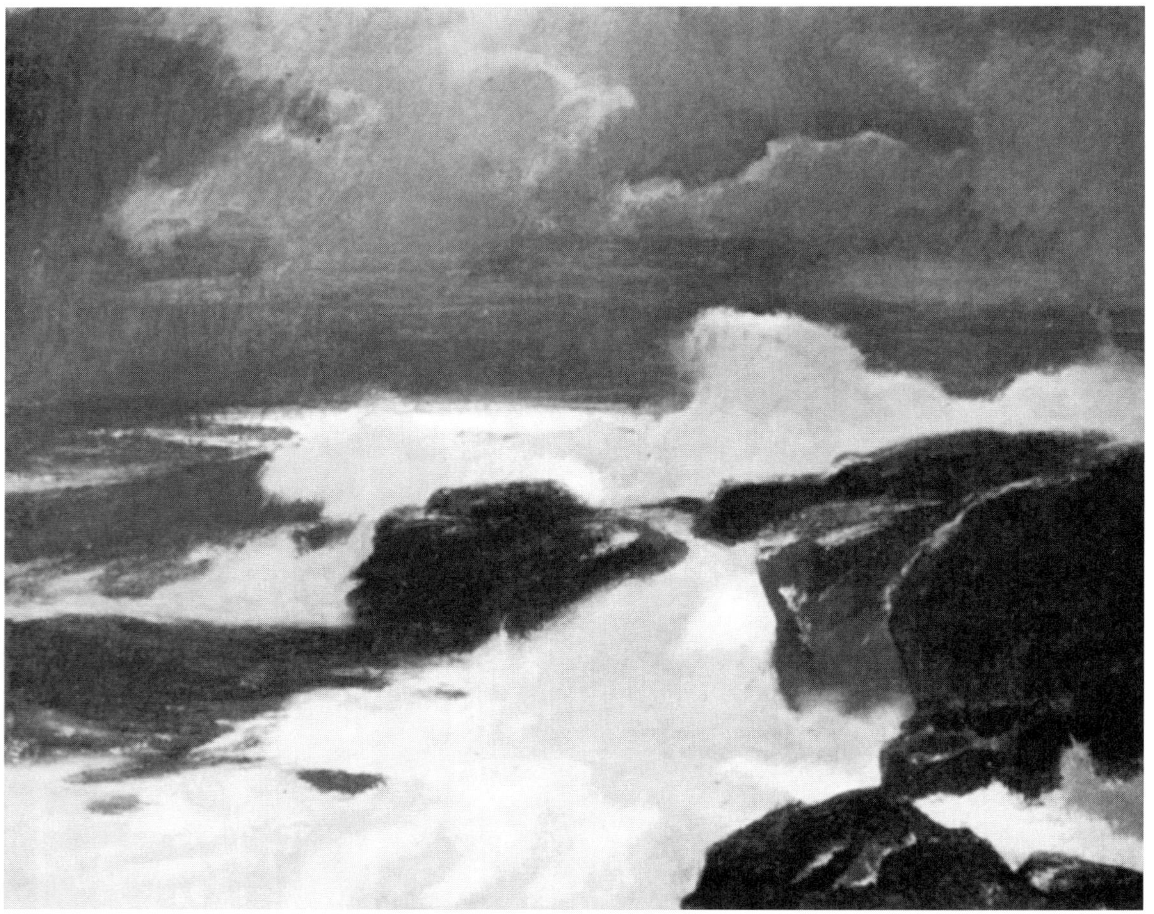

Fig. 17. *Moonlight picture*

Generally, the white foam of the cresting waves and floating foam are part of your light pattern; also the sky and sometimes sunlit rocks or sand. The darks are, of course, the dark sea water and rocks in shadow (Fig. 16A & B) —occasionally the sky, if you are doing a storm or moonlight effect (Fig. 17).

There are a number of types of composition which are used in the black-and-white pattern of most pictures. But before we discuss them, let me briefly explain what is meant by balance in a picture.

Most pictures are composed on the principle of the steelyard balance. If this term is confusing, think of the old idea of the seesaw (Fig. 18). An adult has to sit well in near the center to be balanced by a child out on one end of the seesaw. Applied to a picture, a large mass of either light or dark

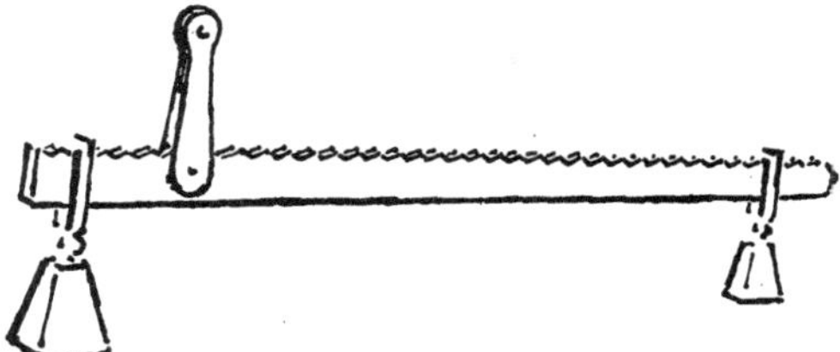

Fig. 18. *Steelyard balance*

near the center of your picture can be balanced by a smaller spot out near the edge (Fig. 19).

A popular type of composition is a light picture with balanced spots of dark. A composition with a dark mass of rocks in the lower right foreground, balanced by rocks out on the *middle* of the left side, would be this type of picture. You could have a light sky, fairly light sea, and white sparkling foam, the rocks being your only real dark spots (Fig. 20).

Many marines are painted with the same composition in reverse. A dark picture with balanced light spots. A moonlight scene with a little light on the water and on some of the foreground foam, with everything else in the picture dark, would fit into this type of composition (Fig. 21).

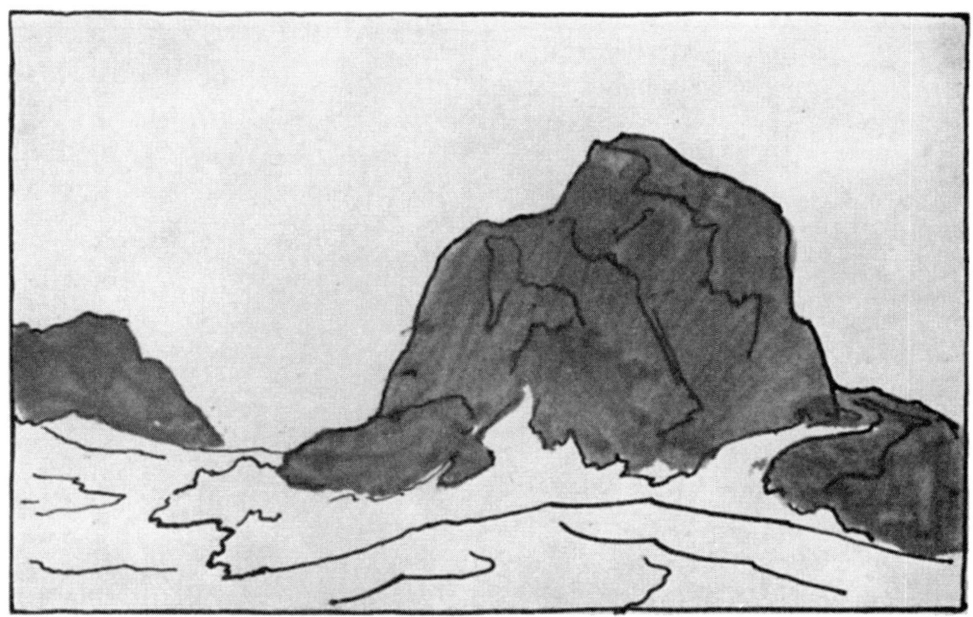

Fig. 19. *Balance of spots. Large spot in center balanced by small one on edge*

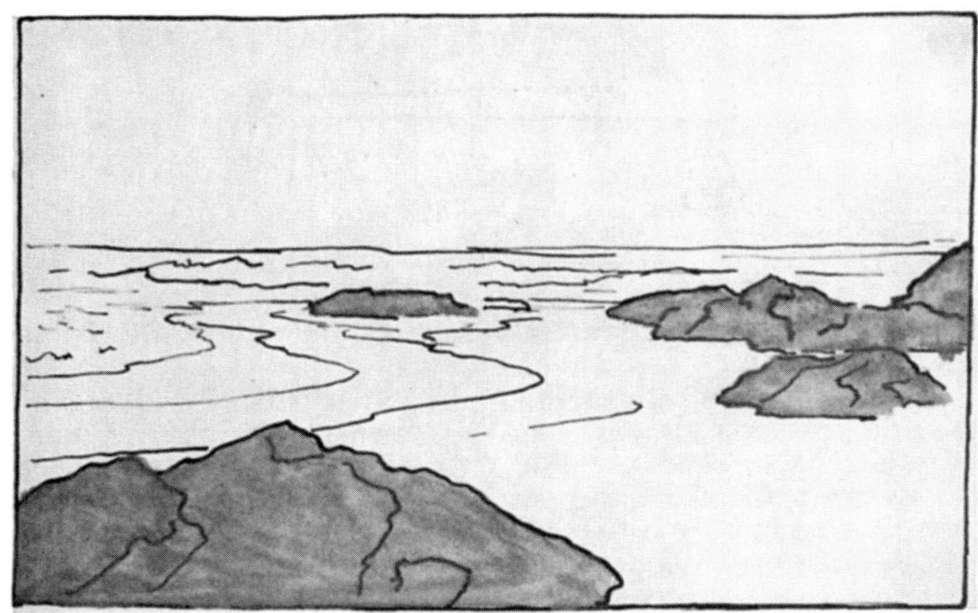

Fig. 20. *Example of balanced spots of dark in a light picture*

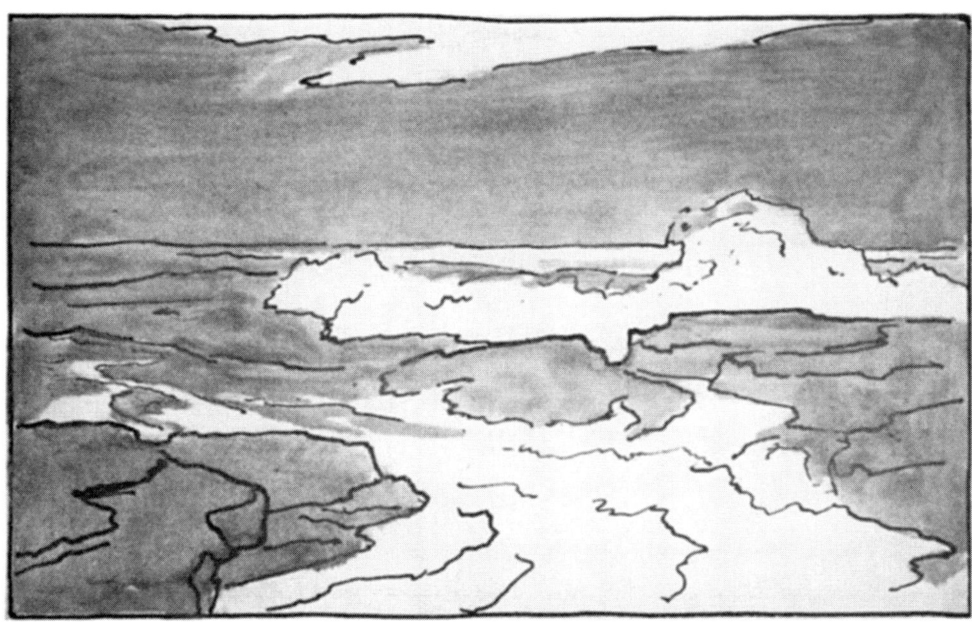

Fig. 21. *Dark picture with balanced spots of light*

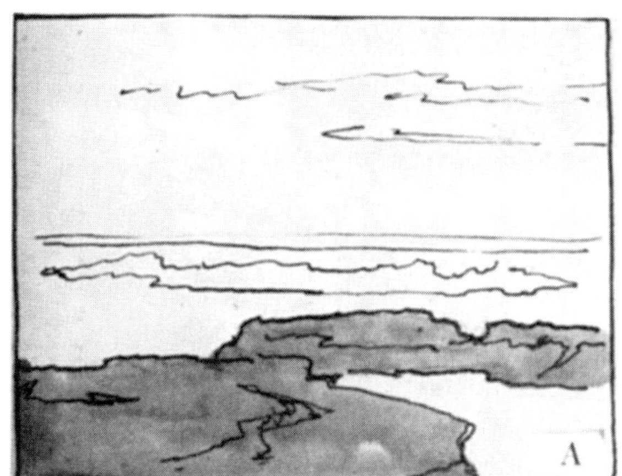

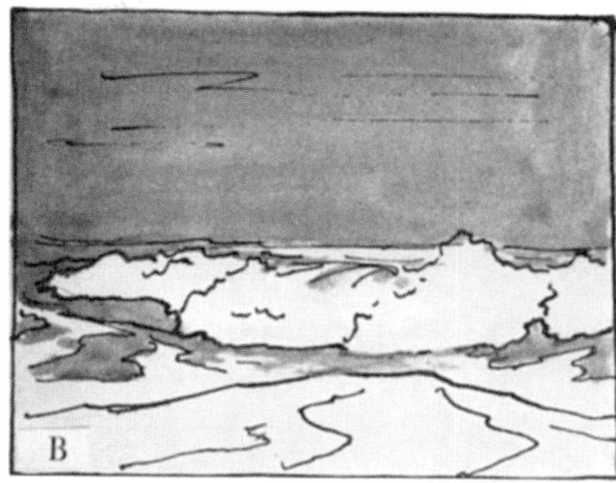


Fig. 22. *Two compositions: (A) dark base, light top (B) same in reverse*

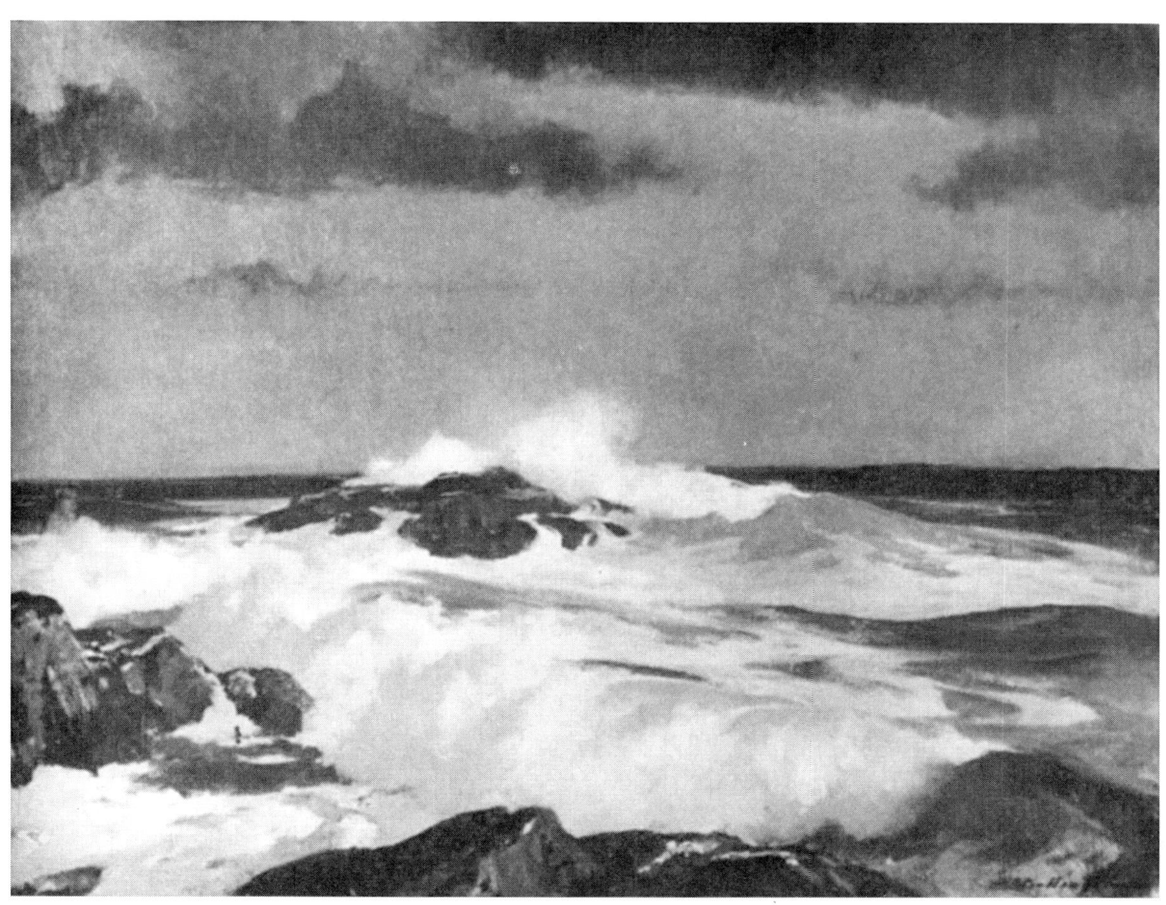

Fig. 23. *Pyramid composition; interest building up the center*

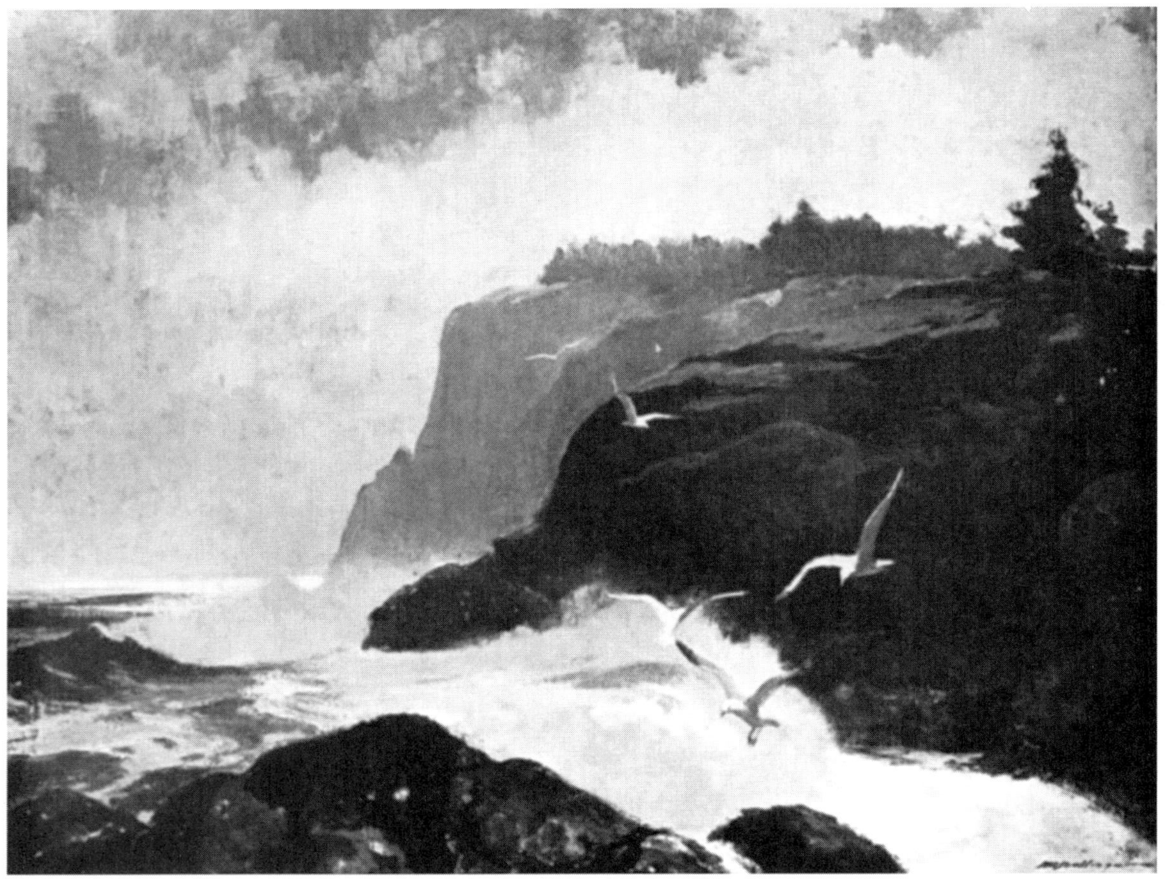

Fig. 24. *L-shaped composition*

A picture with a dark base and a light upper portion is effective; also the same in reverse—dark top, light lower portion (Fig. 22A & B).

A pyramid composition with the weight and interest building up the middle of the canvas is always strong and effective (Fig. 23). There is also the L-shaped composition which builds up one side of the picture (Fig. 24).

I often like to use the S-shaped composition in which the interest swings through the picture from top to bottom like the shape of the letter S. It is always effective (Fig. 25).

Sometimes the picture can be arranged in horizontal bands of light and dark (Fig. 26). Often a combination of several of these compositions can be used in the same picture (Figs. 27 & 28, and Fig. 11).

The difference between painting and drawing is that when you paint you

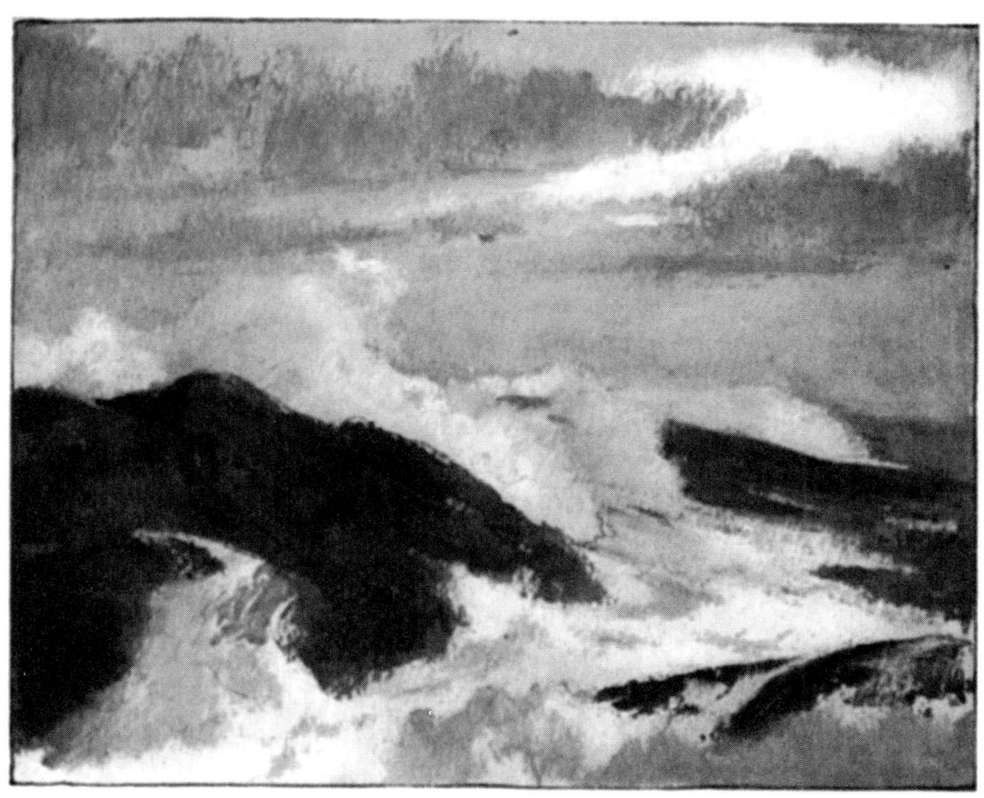

Fig. 25. *S-shaped composition*

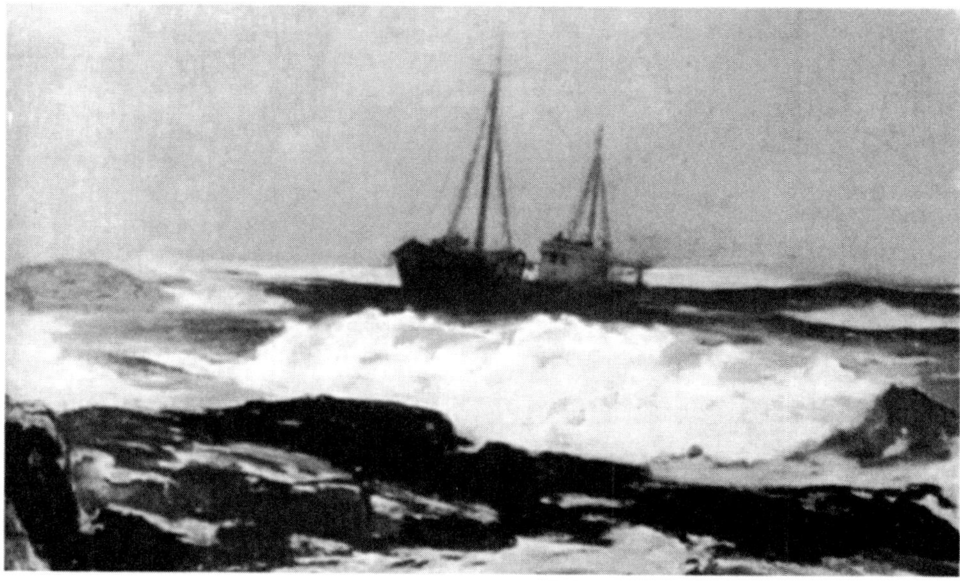

Fig. 26. *Horizontal bands of light and shade*

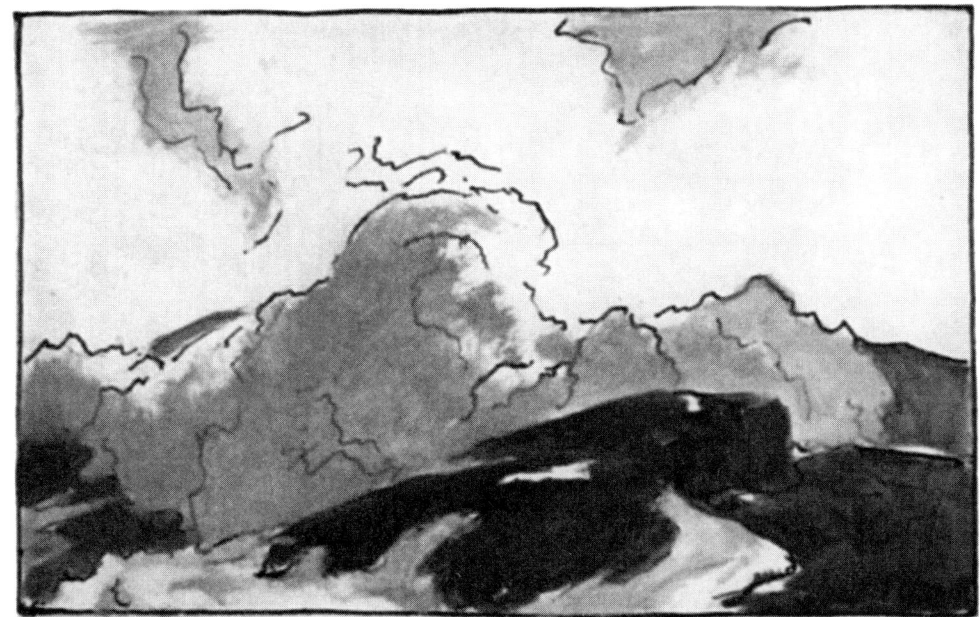

Fig. 27. *Dark base—pyramid composition*

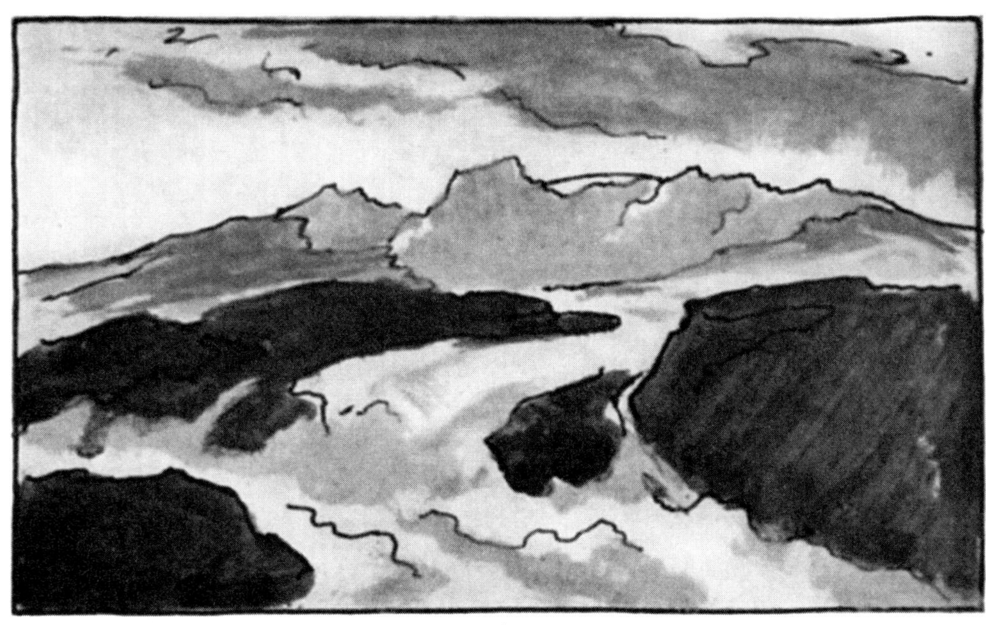

Fig. 28. *Balanced spots—S-shaped composition*

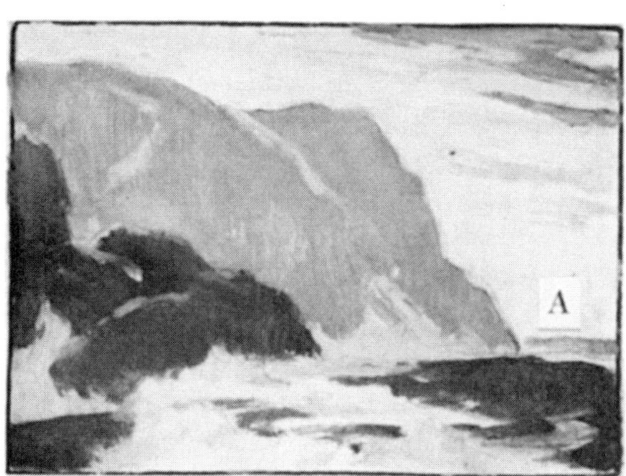


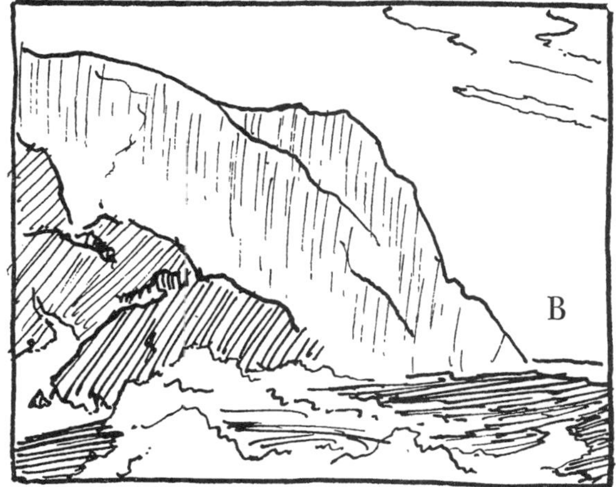


Fig. 29. *Two drawings showing (A) treatment of mass (B) outline style*

work in masses of light and dark which are spots without much outline while a drawing is simply a matter of line and outline, with very few values in it (Fig. 29A & B).

If you are able to see a fine decorative arrangement of your light-and-dark pattern in the picture you wish to paint, you will automatically have a good picture. It will almost paint itself. But if the composition doesn't

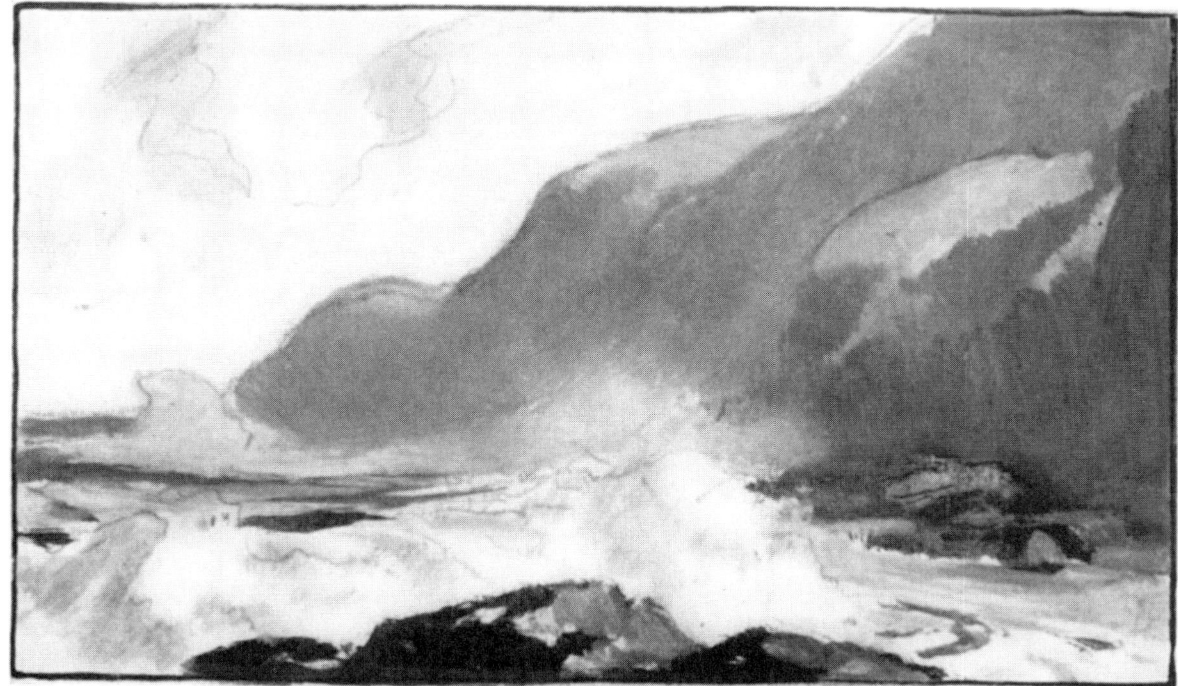

Fig. 30. *Diagram showing all lines vanishing at horizon line*

balance or work out with a decorative design, no matter how much you struggle with it, the picture will never be any good. In that case there is really nothing much left to do except shoot yourself or just give the whole thing up and start another picture somewhere else.

One way to learn composition is to try and figure out the composition used in some of the fine pictures that you like. By studying them you can see how the principles of balance and composition that I have just mentioned can be applied to your own picturemaking.

## SIMPLIFIED PERSPECTIVE

Perspective plays an important part in any picture, so at this point I would like to explain the general idea of perspective.

The receding lines of all objects in a picture converge at a point on the horizon line at the level of your eye. Your eye level determines the horizon line. The eye line and the horizon line are the same. Everything above your eye level comes down, as it recedes, to the height of your eye on the horizon line; everything below comes up to it (Fig. 30). A figure standing on a cliff sees a high horizon line, while a seated figure on a beach will see a low horizon (Fig. 31).

Nearby objects below your eye line seem to tip down and have a more exaggerated perspective, which gradually flattens out as they go away from

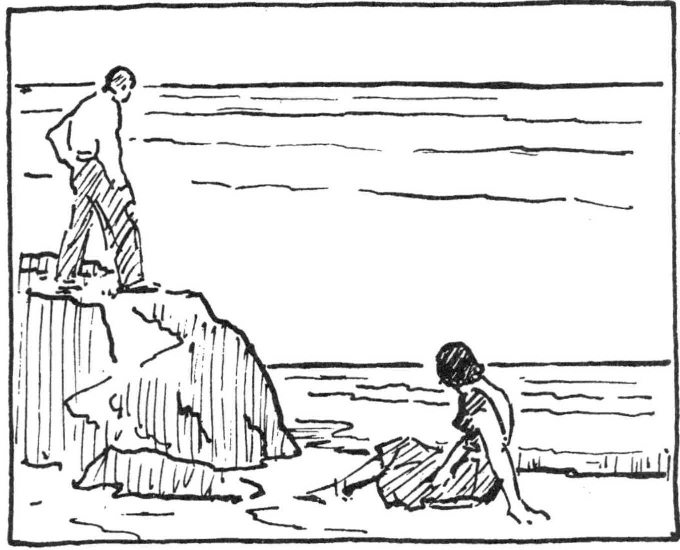

Fig. 31. *Horizon line as seen by figure standing on a cliff and seated figure*

you up towards the horizon line (Fig. 32). In a marine this would mean that the waves would appear closer together and flatter as they near the horizon and, of course, much smaller. The same is true of lines above your eye level. Clouds become smaller and closer together as they near the horizon.

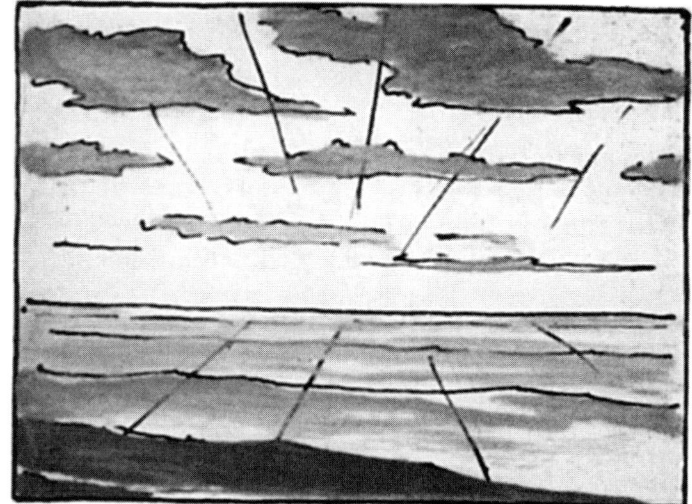

Fig. 32. *Perspective of waves and clouds—flatter near horizon*

An easy way to understand perspective is to hold a thick book horizontally in front of you below your eye line. Note the amount of the top that you see, then gradually bring the book up to the level of your eye and see how the top flattens out until at the eye level the top is only a line (Fig. 33).

There are many books on perspective, but they are mostly so ponderous and complicated that I have never been able to stay awake long enough to finish one. However, I hope you will thoroughly enjoy learning more about this fascinating subject.

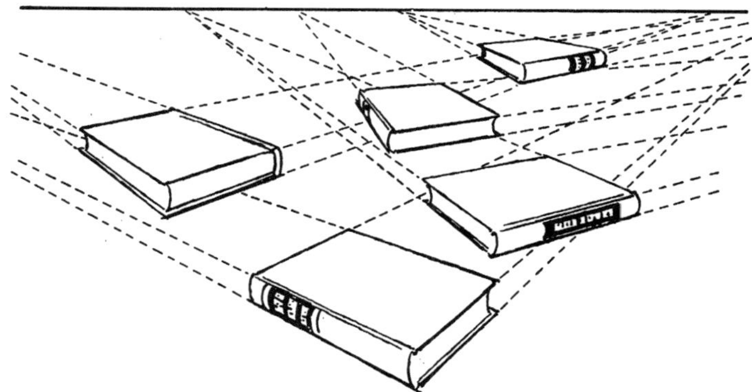

Fig. 33. *Book in perspective: below eye line—three different elevations*

# 4: Planning the Picture

When you go out to the shore to paint, look the scene over carefully and select the first view that looks as though it would make an interesting picture. Don't waste any time trying to find something better, as you may not find it. If the particular scene at which you are looking interests you, that's it.

Next study it from several different directions and try to find the angle that will give you the most pleasing arrangement of the areas of light and shade. Also carefully note the direction of the light and which way the sun is moving, so that your light effect won't change too drastically. It is also a good idea to have a schedule of the tides in the vicinity, so that you can tell whether the tide is rising or falling, and can plan your picture accordingly.

Study the scene for a few minutes and note where the oncoming waves crest; then pick a spot for your center of interest where a wave is really breaking and arrange the rest of the picture to go with it.

It sometimes helps to view the scene through a view finder. This is simply a rectangular opening in a piece of cardboard with a wide enough border around the opening to cut out all but the scene that you are viewing. When you are sketching outdoors, there is so much to see that it is often hard to decide just what to include in your picture and what to leave out. The view finder will help you.

Before starting to paint, it will probably help if you make two or three pencil sketches to be sure that you have a good black-and-white design for your painting.

As the water is the most important part of a seascape, be sure you leave plenty of space for the surf and water; the action of the waves should be the first thing that you plan. The foreground interest with rocks or sandy beach should be only of secondary importance (Fig. 34).

Many students become so interested in drawing the foreground rocks or beach that they put off painting the surf until everything else in the picture is finished; then they find that they haven't any room for the surf.

If, for your center of interest, you decide to have a wave breaking near the middle of your picture, you may perhaps have room for a flattened-out

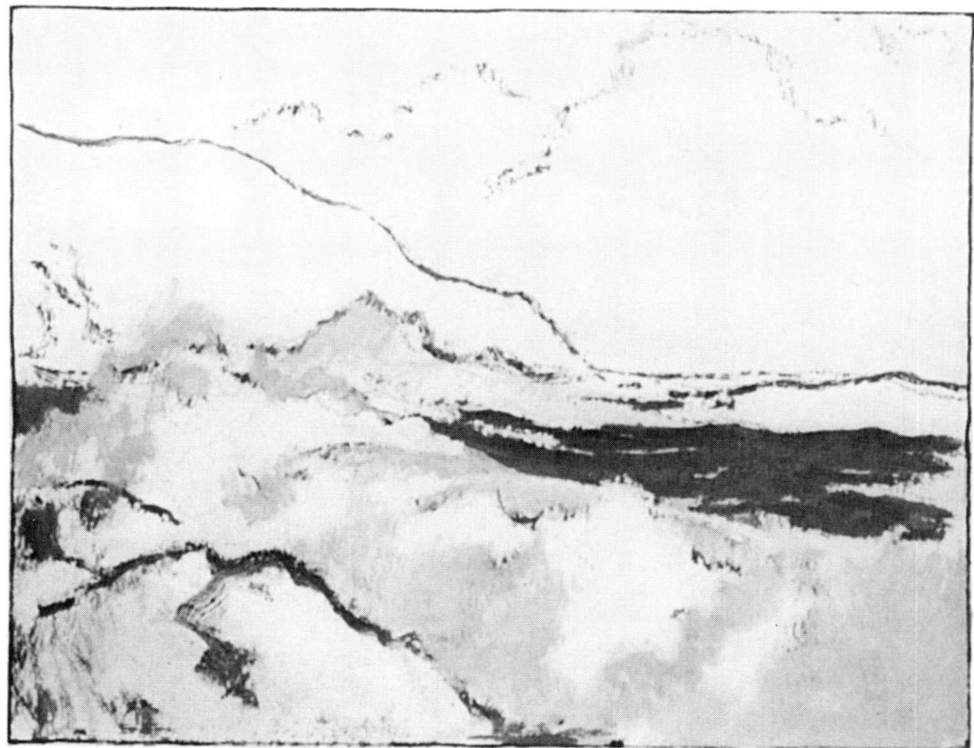

Fig. 34. *Drawing of wave action*

wave pouring over rocks in the foreground. Out at sea you can show other uncrested waves moving in toward shore. A distant headland sometimes lends interest to your pictures (Fig. 35).

Be sure that there is some kind of foreground interest to bring the foreground toward you and to give depth to the picture. You naturally see more detail near you so you must be sure to have something interesting to paint in the foreground.

Rocks in the foreground are always good as they have strong, decorative shapes that make a fine contrast with the light foam of the water. On a sandy beach you can use the dark of the wet sand at the water's edge to give contrast and interest to the foreground; also any interesting details such as tidal pools in the sand, seaweed, driftwood, etc.

It is a good plan when painting a seascape to look along the beach and paint the scene at an angle rather than to look straight out to sea with surf coming right toward you. You can show action more easily by viewing the surf at an angle; you also get away from so many horizontal lines and have a lot more interesting perspective in your picture (Fig. 36).

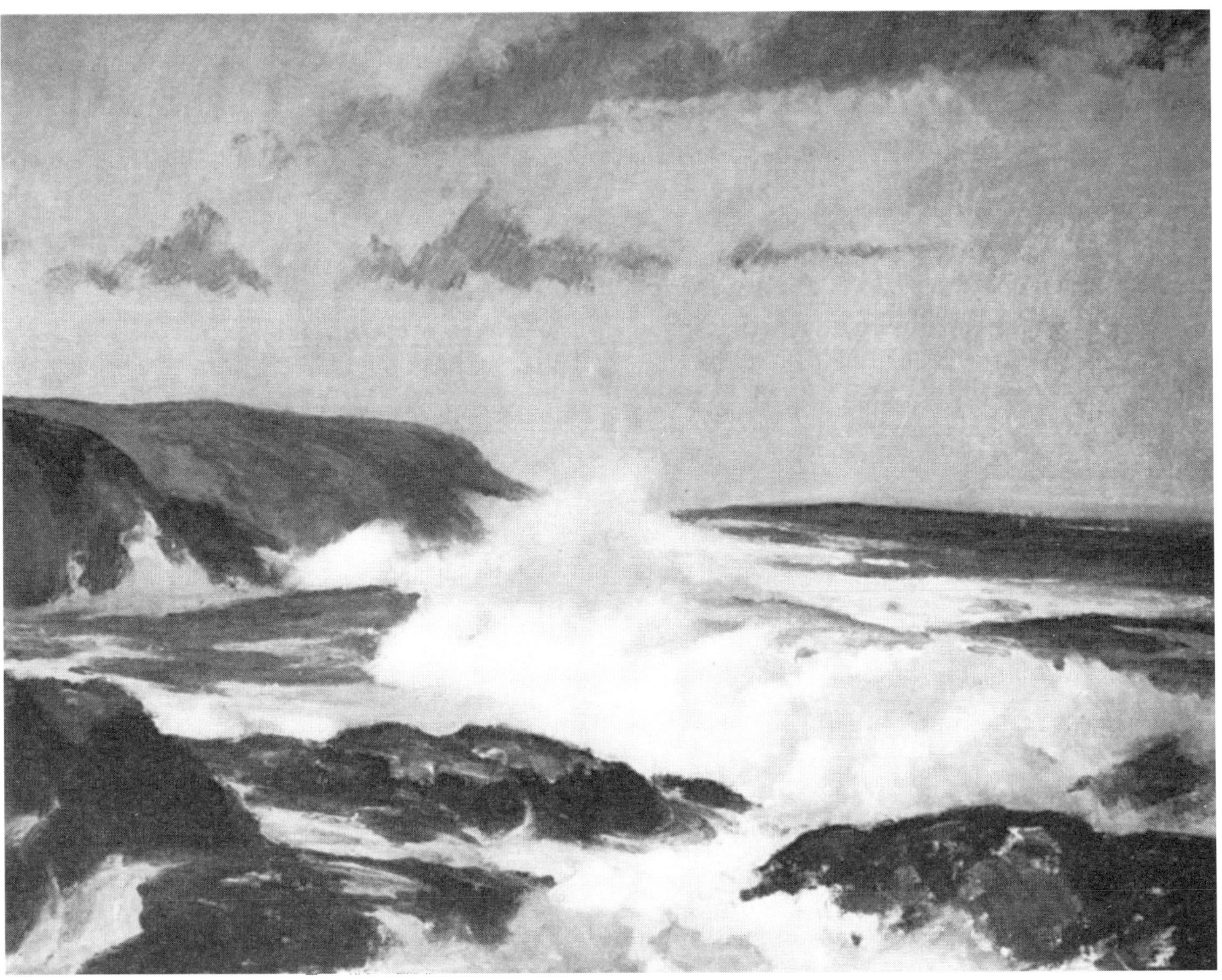

Fig. 35. *Surf on a rocky shore—rocky headland*

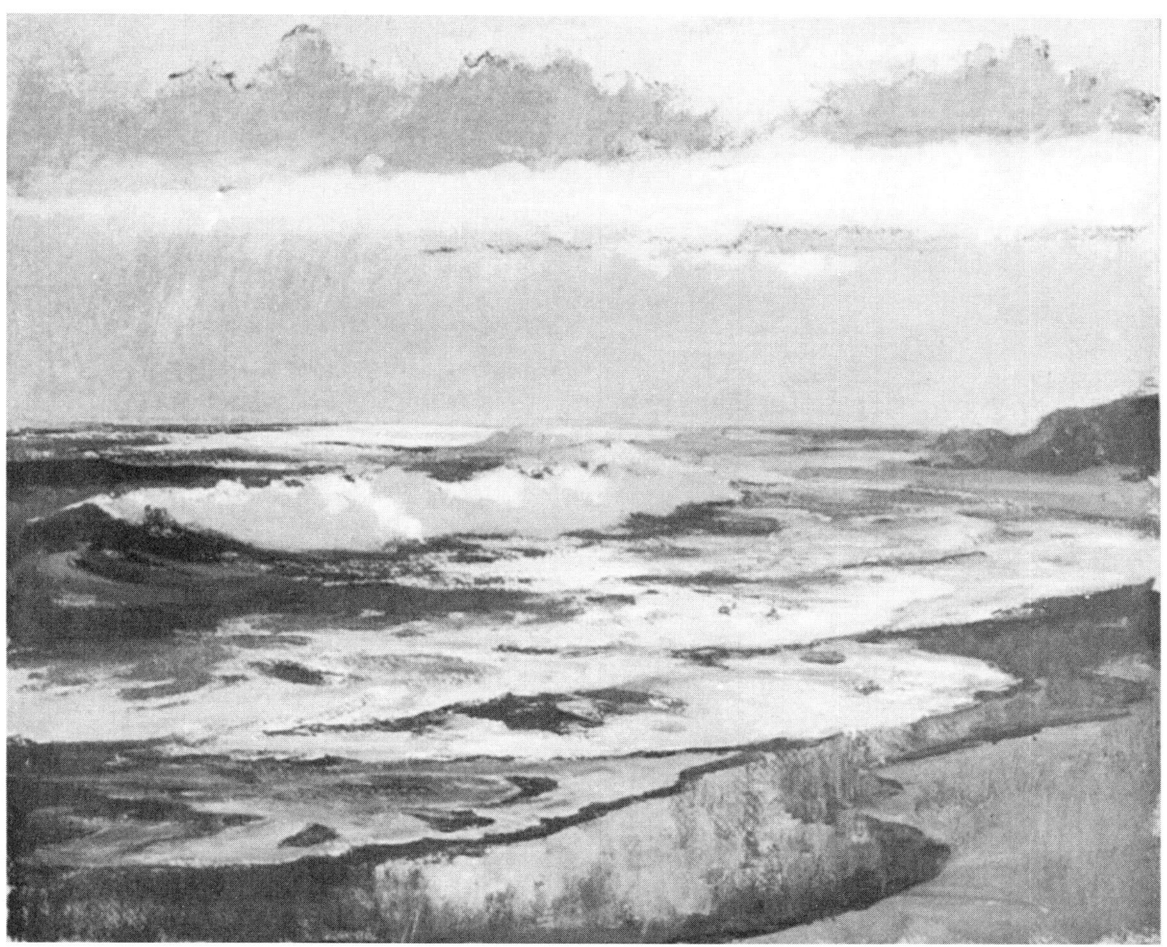

Fig. 36. *Sandy beach painted at an angle*

The direction of the light in your picture is very important. It is a good plan to paint a scene that has a side lighting or one that has the light coming from within the picture, producing a back lighting (Figs. 37 and 38).

Often a flat light over the whole picture is uninteresting and may spoil what might be a fine painting. Sometimes just a simple, diffused down light is effective; that is the kind of light you would get on overcast, foggy, or rainy days (Fig. 39). You can't paint a picture unless you know exactly where the light is coming from; so be sure to figure out the direction of your light before you start to paint.

To summarize—when planning the picture, first find a view that appeals to you; then select the best painting angle from a composition standpoint. Familiarize yourself with the movement of the tides and be sure that the

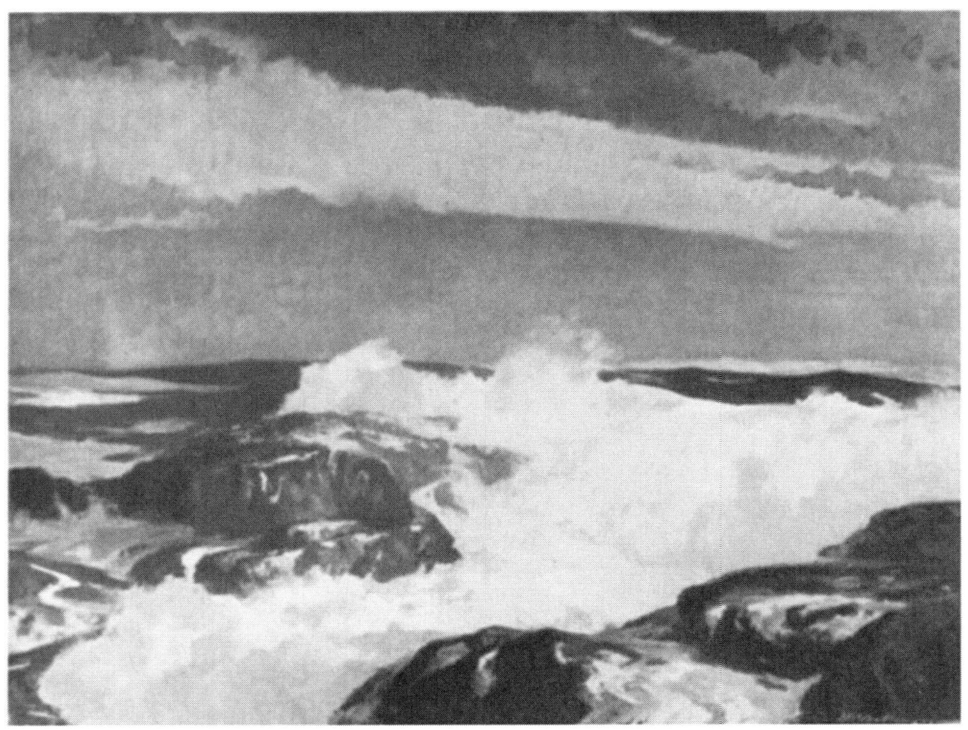

Fig. 37. *Picture with side lighting*

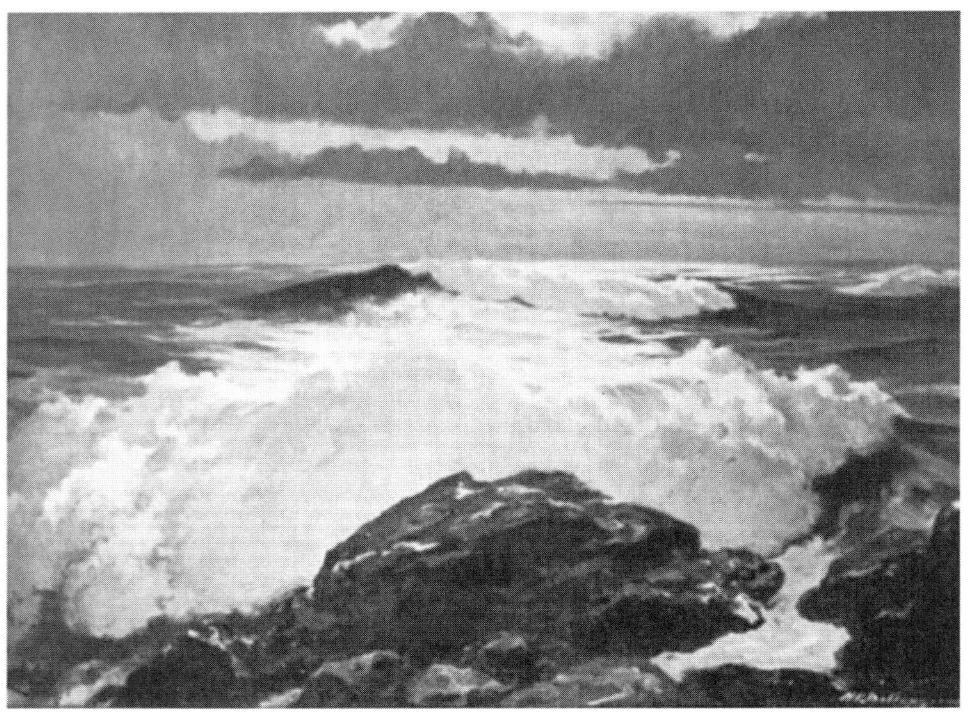

Fig. 38. *Back lighting*

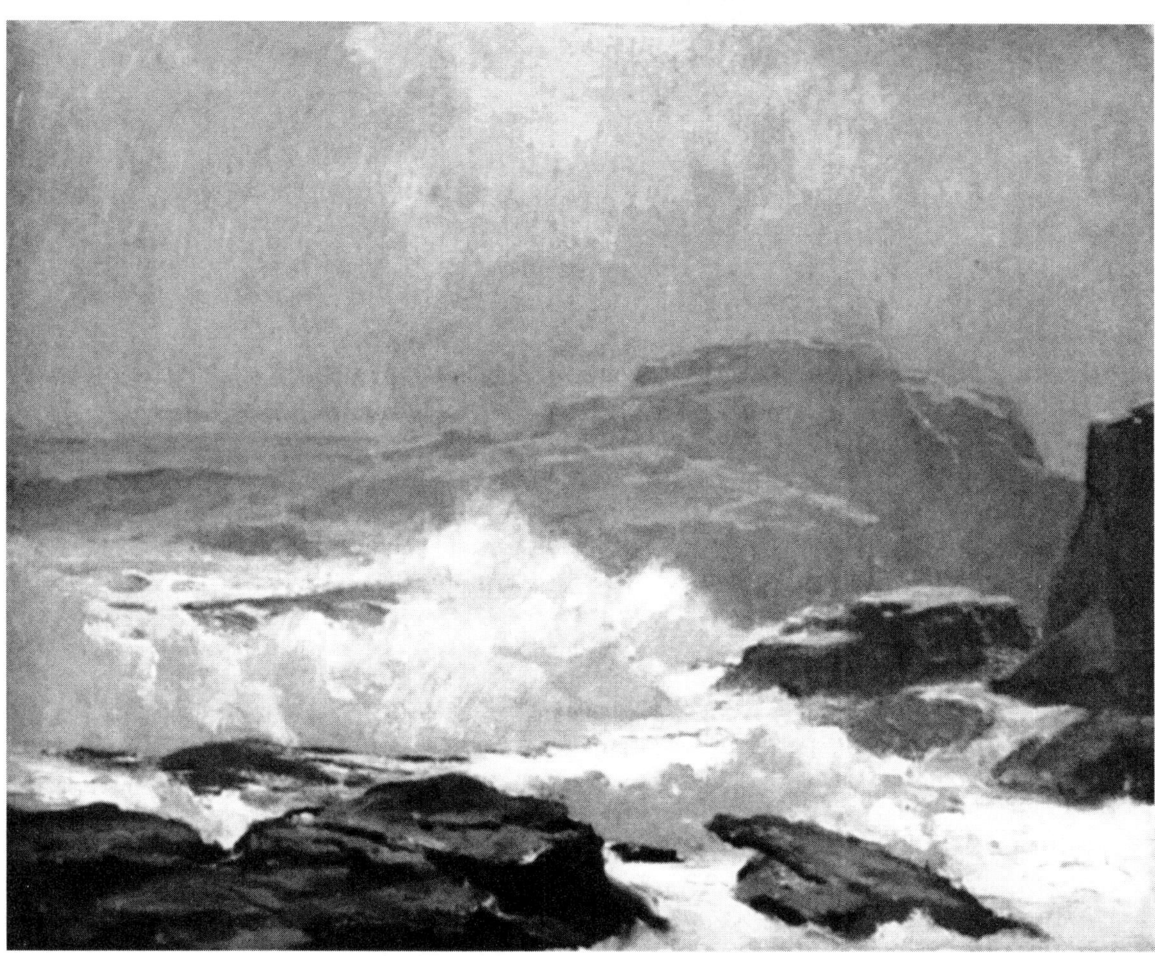

Fig. 39. *Diffused down lighting*

direction of the light will give you enough light and shade with which to work.

Painting by the sea is a tough job, as the ocean isn't at all cooperative and sometimes seems violently antagonistic. However, it is surely worth the trouble if you are able to capture some of the grandeur and beauty of the sea in your canvas. So think nothing of it if your canvas blows away or you get washed off a rock. To the marine painter, this is all good, clean fun.

# 5: Color Mixing—The Ballinger Palette

The choice of colors that I use is simply the standard high-keyed palette that has been in general use for years. As I have previously explained, it contains no earth colors, greens, or black.

I believe that a simple palette with only the colors that you absolutely need is better than a palette with every conceivable color on it. When, for example, you mix a green with cerulean blue and cadmium yellow pale, you get more sparkle and vibration from a combination of the two colors than you would from a green that comes right out of the tube.

However, it takes practice and study to match in color and value the objects that you wish to paint. I suggest that you could profitably spend a few hours just trying to match with your paints the color and value of everything in the living room. Try to match the color of the ceiling, walls, floor, draperies, and furniture. Match the general color of the carpet, for instance, both in the light and in the shadows. Try the same idea outdoors; no drawing, just matching the colors you see before you. You will find that

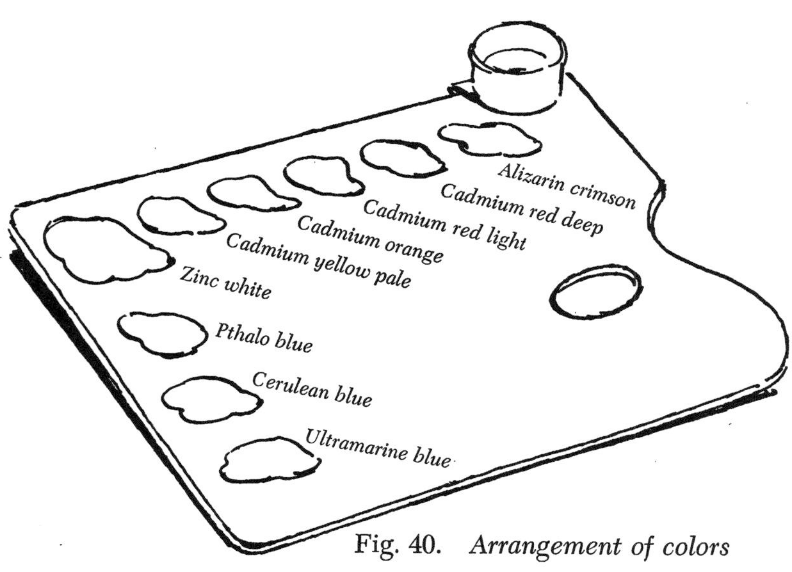


Fig. 40. *Arrangement of colors*

if you can match a color you see in front of you, you will be well on the way to painting a picture. Most beginners can identify a color in front of them all right, but are unable to combine the colors on their palette in the proper proportions to produce the desired color. One thing to remember—go gently when you are trying to mix a color. Sometimes just a touch of color on the corner of a brush will be all you need. While it does take a lot of practice to match the exact color and tone that you see before you, still it can be done with perseverance.

The colors on your palette start with the cool ones on the left; then white; then your warm colors, arranged in the order of the color wheel (Fig. 40).

1. *Ultramarine blue.* This is a dark purplish blue which, combined with red, can be used for your strong darks. I use the blue to darken the color, then add cadmium red deep, cadmium red light or sometimes cadmium orange, depending on how warm a dark I desire. As a rule it is better to have your darks a little on the warm side.

2. *Cerulean blue.* This is a paler, more greenish blue and makes lovely pearly grays when combined with cadmium red light and white. If you desire more of a tan color, add a little cadmium orange to your gray mixture. However, don't add cadmium yellow pale, as this will immediately produce a greenish tint. Cerulean blue is a very useful color, as it is already grayed-up to start with as it comes from the tube, and can be used to gray-up other colors or combinations of colors. It makes fine light greens when combined with cadmium yellow pale.

3. *Phthalo blue.* Once in a while you can use a little phthalo blue or viridian for some special spot, but I don't think that you will often need it.

4. *Zinc white.* This should be fairly liquid, so that it will mix readily with your colors and flow on easily.

5. *Cadmium yellow pale* is like a chrome yellow or lemon yellow. Be sure that they don't sell you cadmium yellow light, as this is warmer and doesn't work as well.

6. *Cadmium orange* is just about the warmest kind of a yellow. By combining varying quantities of cadmium yellow pale and cadmium orange, you can get any degree of warmth that you desire in your yellows.

7. *Cadmium red light.* Be sure that it is an orange red like a vermilion.

8. *Cadmium red deep or medium.* Either of these is all right for use in your dark reds or combined with ultramarine blue to make a black.

9. *Alizarin crimson.* I prefer this to rose madder, as it is a more permanent color. Only occasionally do I use alizarin crimson in a picture, as it is liable to turn any color with which it is mixed into an unpleasant purple. I don't like too much purple in a picture as it makes a painting look like a cheap lithograph. However, there is one time that alizarin crimson is very effective.

When combined with cadmium red light, it makes a strong and brilliant red.

Here in brief is the way to mix some of the colors that you will use most often:

*Black.* Ultramarine blue and cadmium red deep with just a touch of cadmium orange will give the effect of black paint, only, it will have more brilliance than black paint which generally resembles shoe polish.

*Darks.* Use ultramarine blue and cadmium red deep or cadmium red light to warm it up a little. You should use more red and less blue than if you were trying to mix a black. Don't mix alizarin crimson with ultramarine blue unless you want a rip-roaring purple.

*Grays.* Cerulean blue and cadmium red light with white.

*Tans.* Add a little orange to the above gray mixture.

*Greens.* For light greens, cerulean blue and cadmium yellow pale. For darker greens, use ultramarine blue instead of cerulean. If you want a dark olive green, use ultramarine blue and cadmium orange, sometimes combined with white.

Try not to mix your paint too much on the palette. You will have more color vibration when you apply your paint to the canvas if it isn't overly blended. You will get the effect of broken color which will give sparkle to the picture.

As a rule, when you finish a day's painting you clean off only the center of your palette, leaving the gobs of color around the edges. They will usually stay fresh for several days, so the paint won't be wasted.

If you use a disposable paper palette, you can transfer the piles of paint to a fresh sheet and throw the old one away.

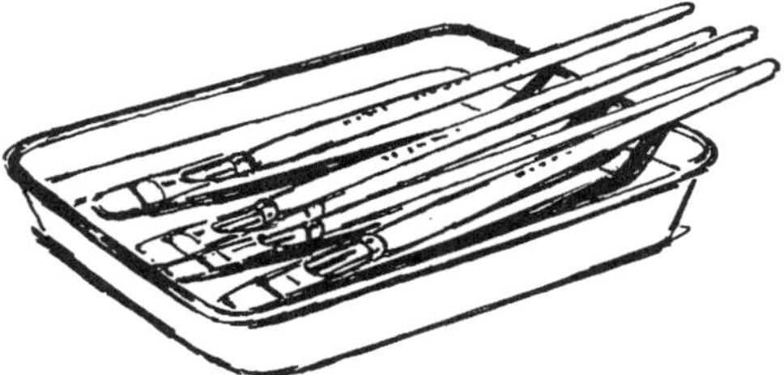

Fig. 41. *Brushes in flat tin of kerosene*

I always place my brushes in a flat tin of kerosene when I finish the day's painting (Fig. 41). When you want to use them again, just wipe them off thoroughly with a rag and they will be ready to use. Every two or three weeks you can give them a real bath with brown laundry soap and water, taking care to rub the soap into the bristles right up to the ferrule of the brush so that no paint will harden in the bristles to spoil your brushes.

# COLOR PLATES

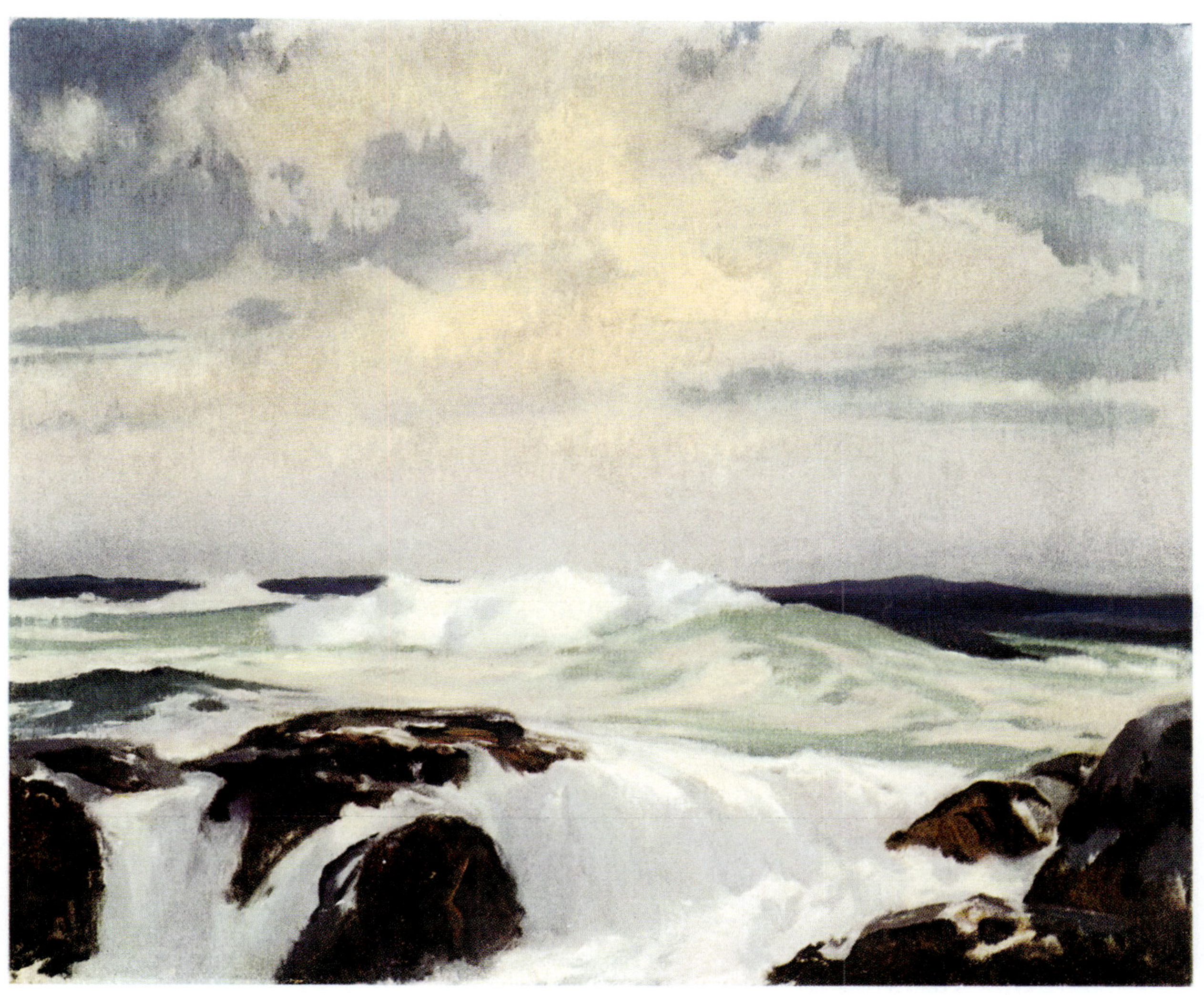

*Cloud pattern in harmony with sea movement, oil, 16 x 20*

Plate 1. Painting Skies

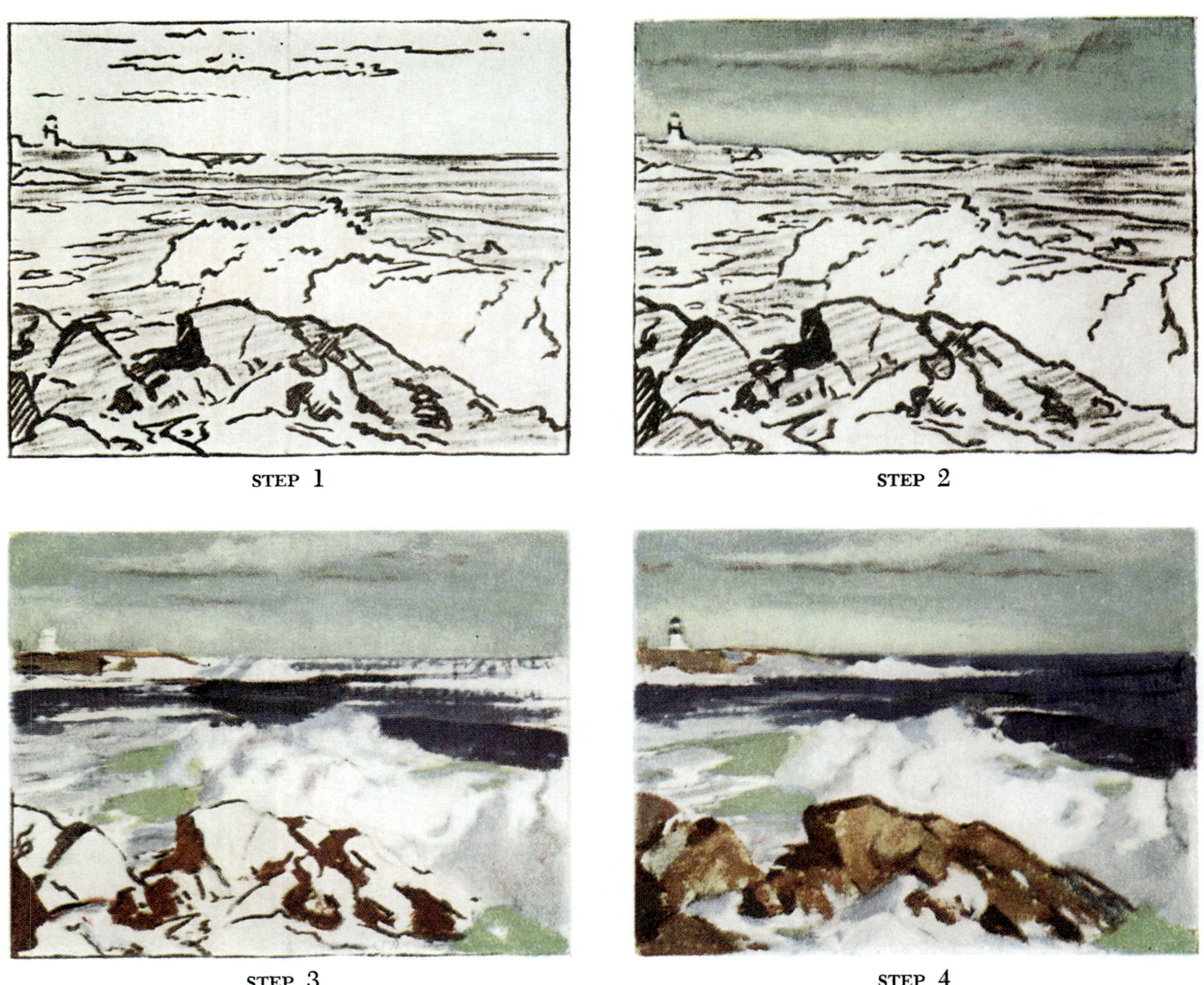


Plate 2. Painting Surf on a Rocky Shore

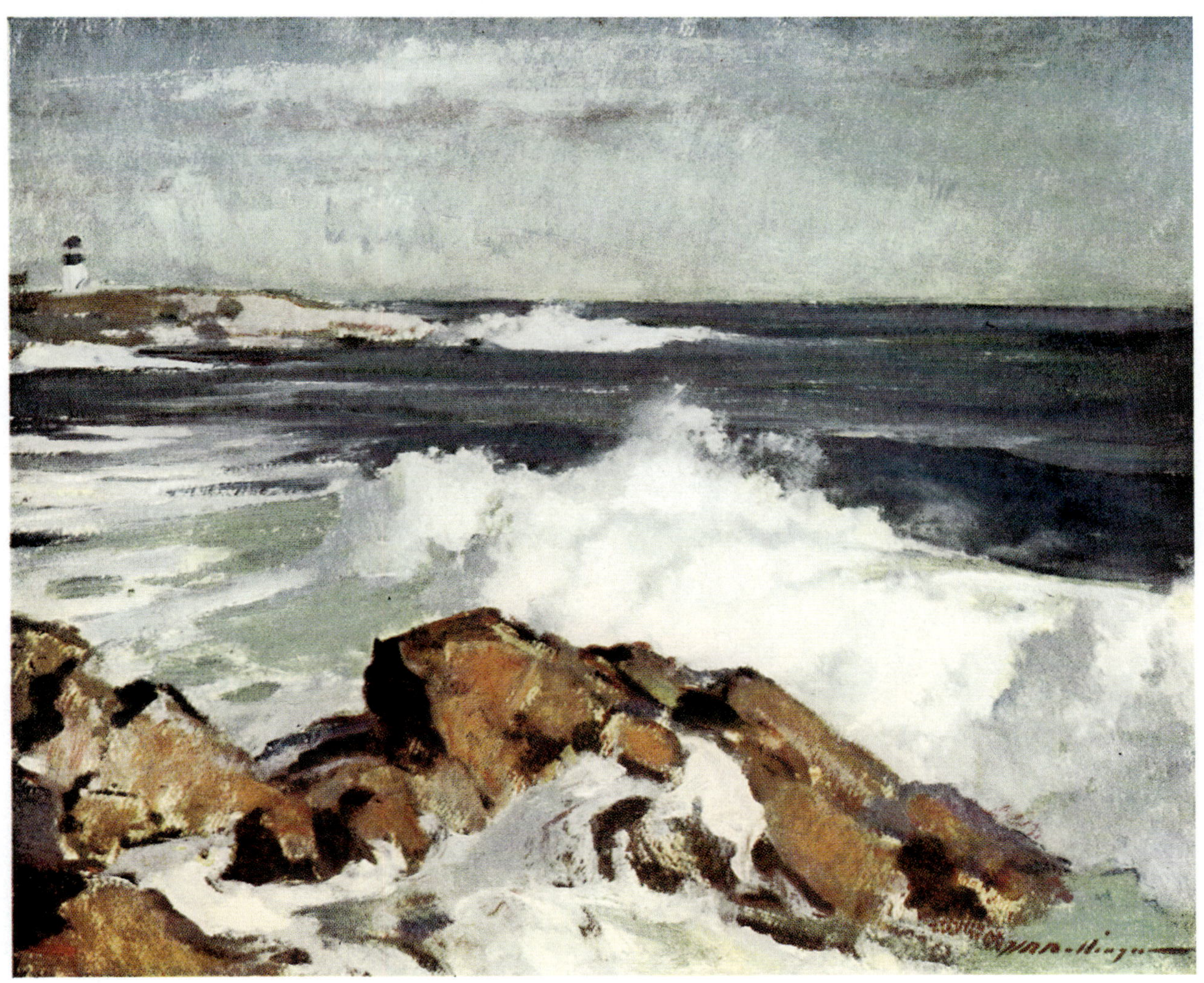

*The completed demonstration, oil, 16 x 20*

Plate 3. Painting Surf on a Rocky Shore

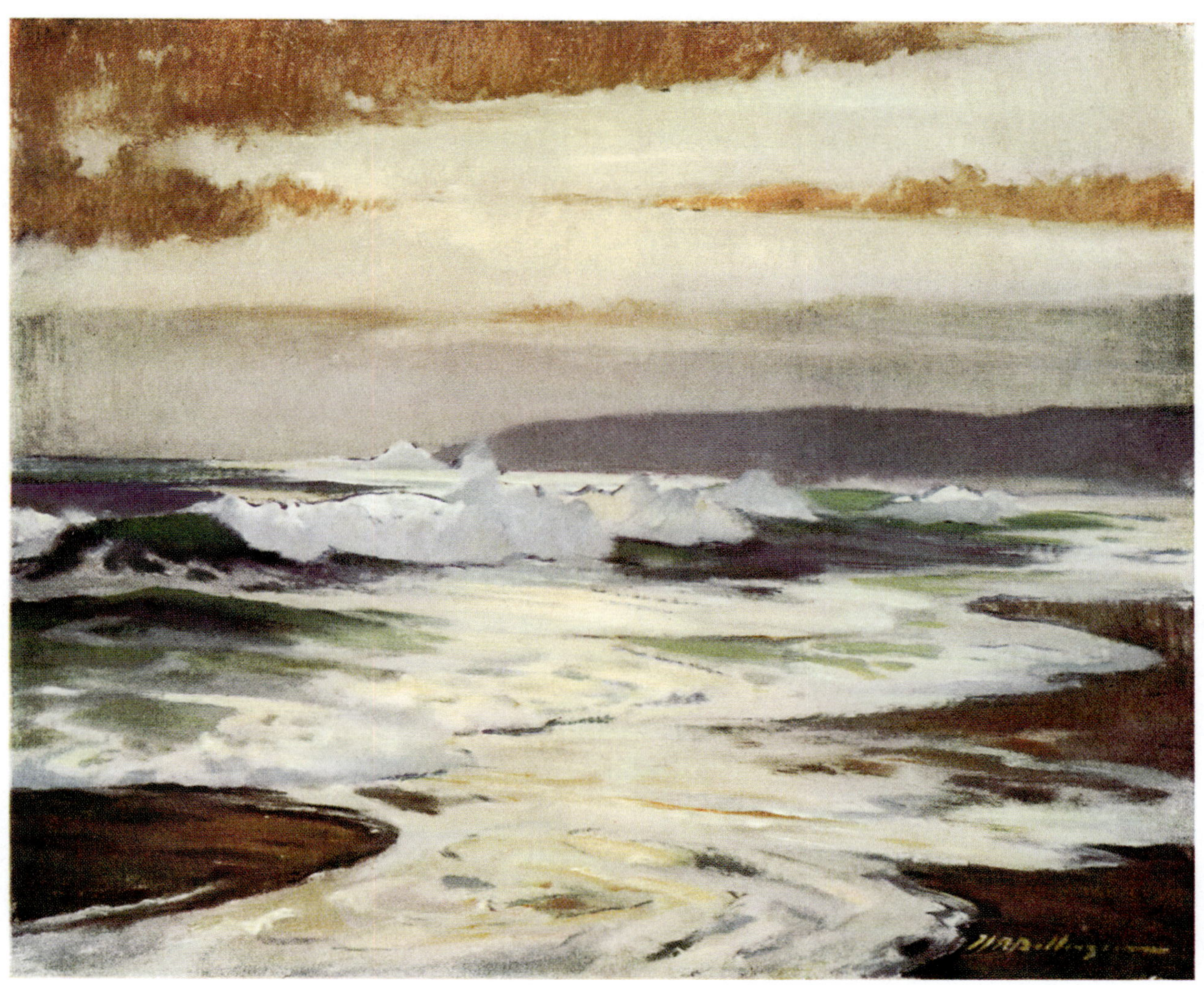

*Demonstration of painting running surf on a sandy beach, oil, 16 x 20*

Plate 4. Painting Surf on a Sandy Beach

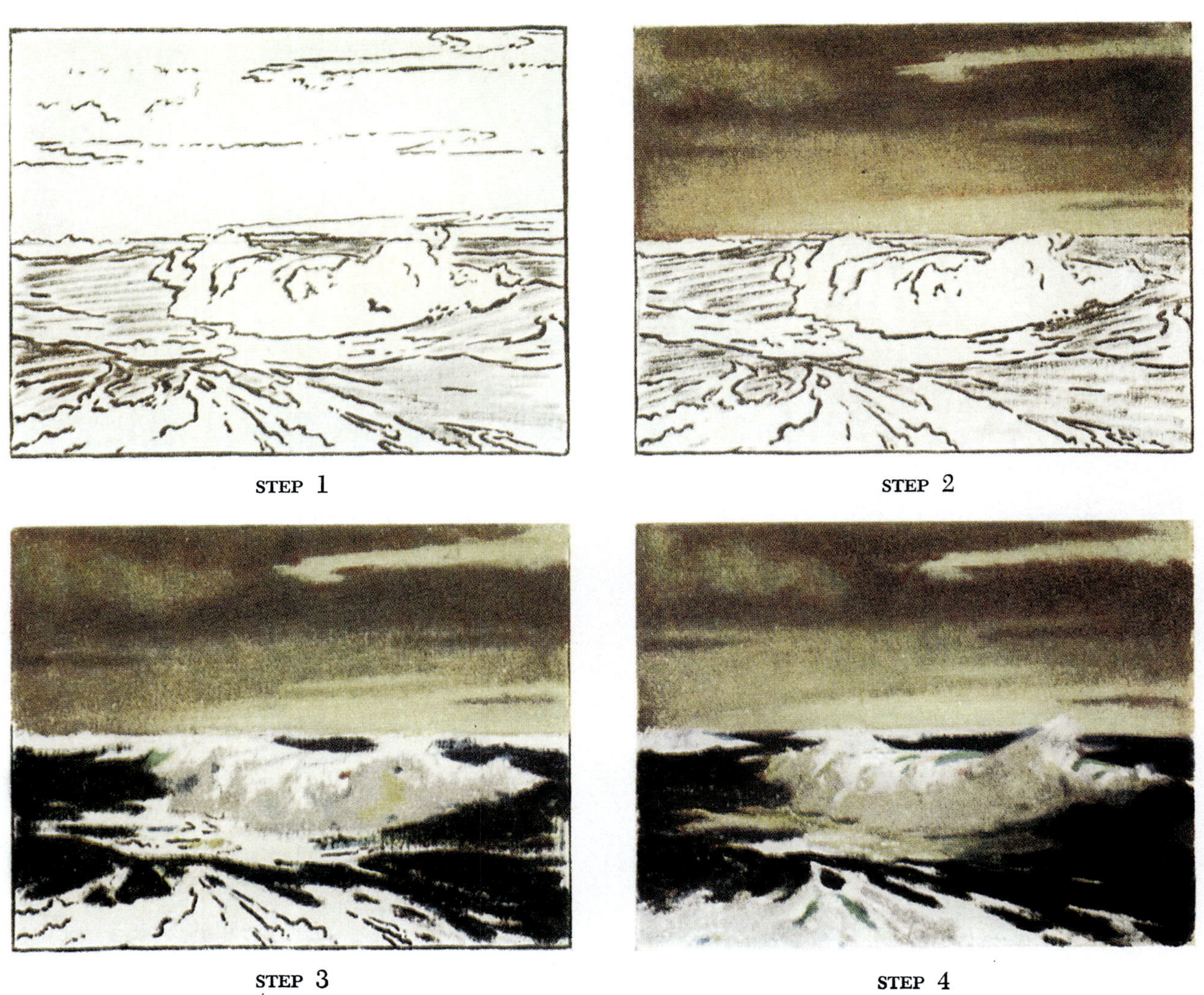

STEP 1

STEP 2

STEP 3

STEP 4

Plate 5. Open Sea Painting

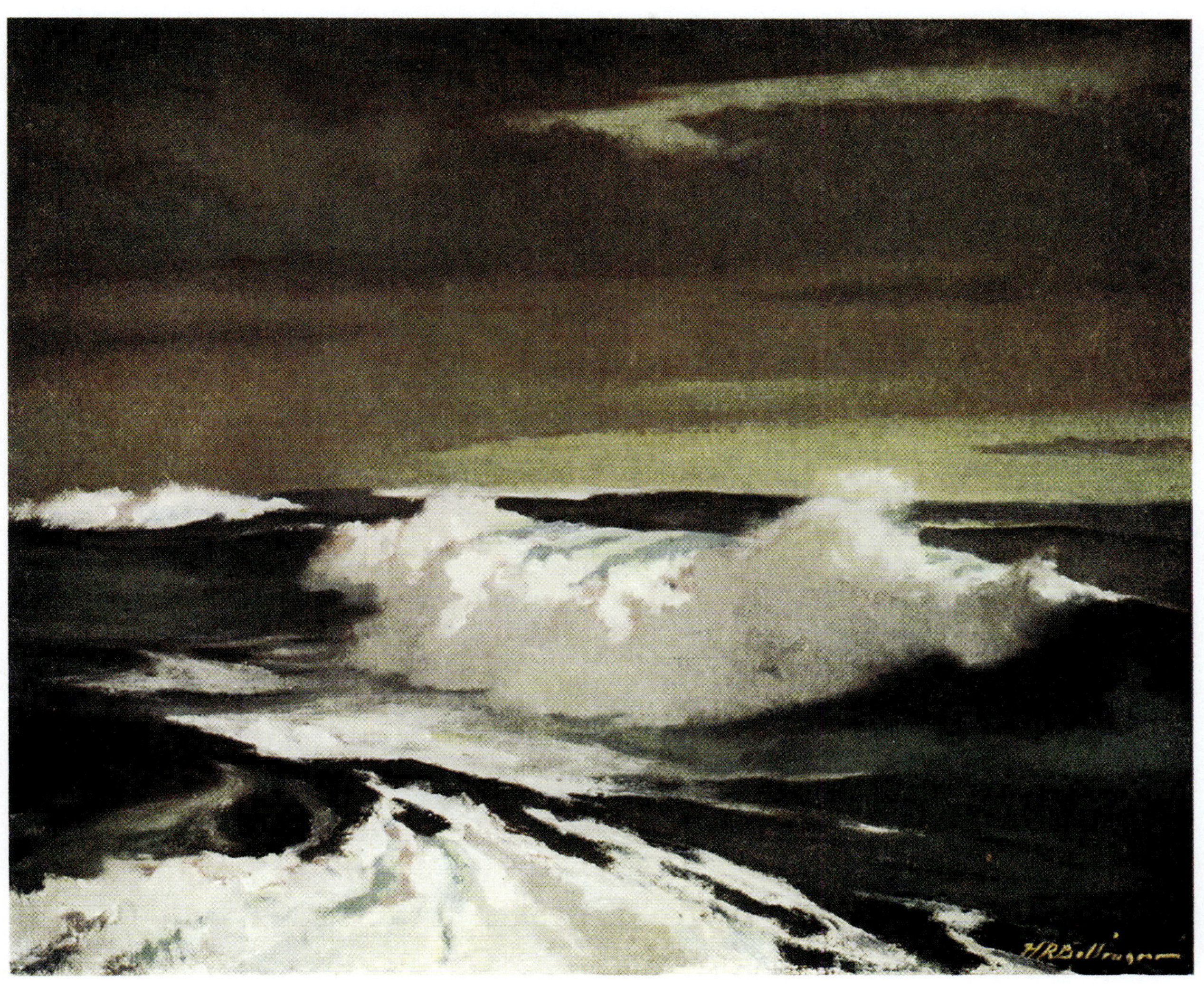

*A simple dark and light pattern broadly painted, oil, 16 x 20*

## Plate 6. Open Sea Painting

# 6: Painting the Picture

A marine is really a portrait of one nice big wave, with plenty of splash to it. Everything else in the picture should contribute to the importance of the wave which is your center of interest.

Sometimes you will have your important wave surging over the foreground rocks or you will feature the water pouring off rocks further out to sea after the wave has rolled into shore. But mostly you will feature a big wave cresting near the center of the picture and dashing against some centrally located rocks. A good moment to paint your important wave is just as the sheet of water at the top of the wave is disappearing and you have a large impressive mass of white foam tumbling down the front of the wave toward you.

As I mentioned in the preceding chapter, first pick out the spot that you wish to paint; study the scene carefully and note where the waves are really breaking. Next, decide on the action that will look best in your picture. Note the direction of the light and whether the tide is rising or falling. Then, decide just how much of the scene to include in your picture. It is a good plan to concentrate on the most interesting portion of the scene before you, rather than to try to paint a wide panorama. Generally there is too much to see in the scene before you and you have enough information for several pictures. The view finder, as previously mentioned, often helps in deciding what portion of the scene to paint, as it automatically eliminates all of the scene except the particular part that you are studying (Fig. 42).

When starting to paint the picture, carefully draw in outline, with a little blue paint and lots of medium, the shapes of all the important masses of light and dark in the picture. The pattern of your picture will thus be clearly defined. Be sure to draw the shapes of the white foam areas of your cresting waves and floating foam, as well as the shapes of the rocks and any of the cloud formations in the sky.

When you have decided on the black-and-white design of your picture, then you must study the color scheme of the scene in front of you.

So far in this book, I have talked more about values and the light and dark pattern of a picture than I have about color. I have done this because I consider that tone and value are the very framework of your picture. If

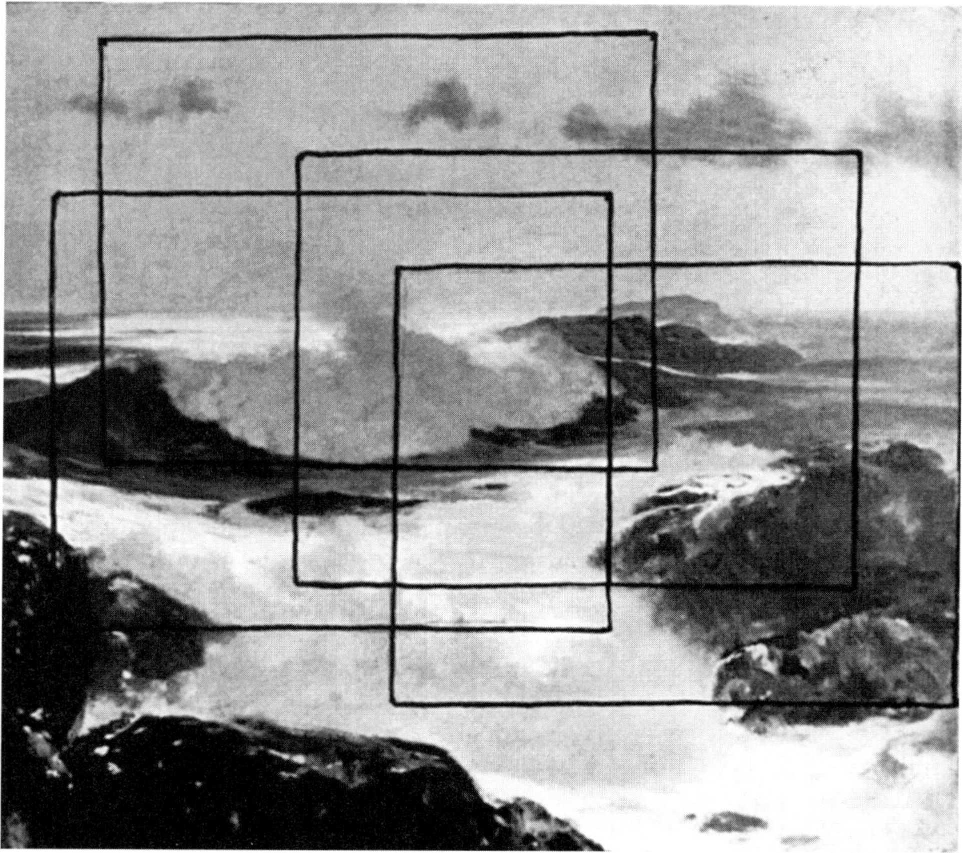

Fig. 42. *Use of view finder*

you haven't a good black-and-white pattern to your picture, you haven't a good picture. However, color is extremely important and should always be carefully considered.

It is a fine idea to have imagination and to be able to change everything around in your picture to suit yourself. But I think you will learn more about marine painting if at first you honestly try to paint the scene before you just exactly the way it looks. By painting just what you see in a literal manner, you gain a lot of valuable information about the surf and shore that will be a great help to you later on when you want to paint a picture in the studio. After all, you can't even start to paint a picture until you have learned to match on canvas the exact color and value of the object that you are trying to paint.

Most of the colors in nature are a combination of two or more of the colors on your palette and are slightly grayed up; in other words, not as brilliant as the raw color that comes straight out of the tube.

At first you may find it difficult to mix the grayed-up colors that you see before you. However, with practice you will learn just which colors will produce the desired results.

All color can be roughly divided into warm or cool color. So when you look at a scene in nature, you must first decide whether the area that you wish to paint is on the warm or the cool side of your palette. When you have once decided on that point, you will find it a lot easier to match the toned-down color that you see before you. For instance, most shadows are on the warm side, except when cooled off by the reflected blue of the sky. A shadow cast on a sandy beach from a boat pulled up on the shore would be cool where the blue sky could reflect down onto it, but would be warm under the boat where there would be no reflection from the sky.

Shadows in rocks are quite warm except where a little of the cool color of the sky hits down on some of the top planes in shadow.

Sunlight isn't quite as yellow or pink as you would expect it to be. When painting it, be sure that it has quite a lot of white in it and only a delicate amount of warmth.

The only time that sunlight is really warm is just before the sunset when the sunlight has a strong orange-pink color.

Sea water is a greenish blue close by, when you look directly into it; it becomes a more grayed-up blue in the distance where it picks up the color of the sky.

When painting the foam of a cresting wave, the part in sunlight will be a little warm, while the shadows will be on the cool side.

When you are ready to start the actual painting, begin first with the sky, as it sets the key for the whole picture. The sky color reflects down into all horizontal planes. If you have a stormy sky your whole picture will have to be a stormy one, with the gray of the sky reflecting down onto all of the top planes of the picture. Sunny skies produce sunny pictures, with warm pinks and yellows throughout the picture. When starting the sky you can paint rather thinly and without much finish, but at least cover the canvas with approximately the right color and value. Indicate any cloud formations at the same time.

The sky in a marine should be toned down just a little, so that the white foam or the glitter on the water will be the lightest spot in the picture.

Next, start right in painting the water. Begin with your center of interest, which will probably be a cresting wave near the middle of your canvas. Paint the foam of the breaking wave and the water around it. Also any adjacent rocks that are influencing the action of the wave.

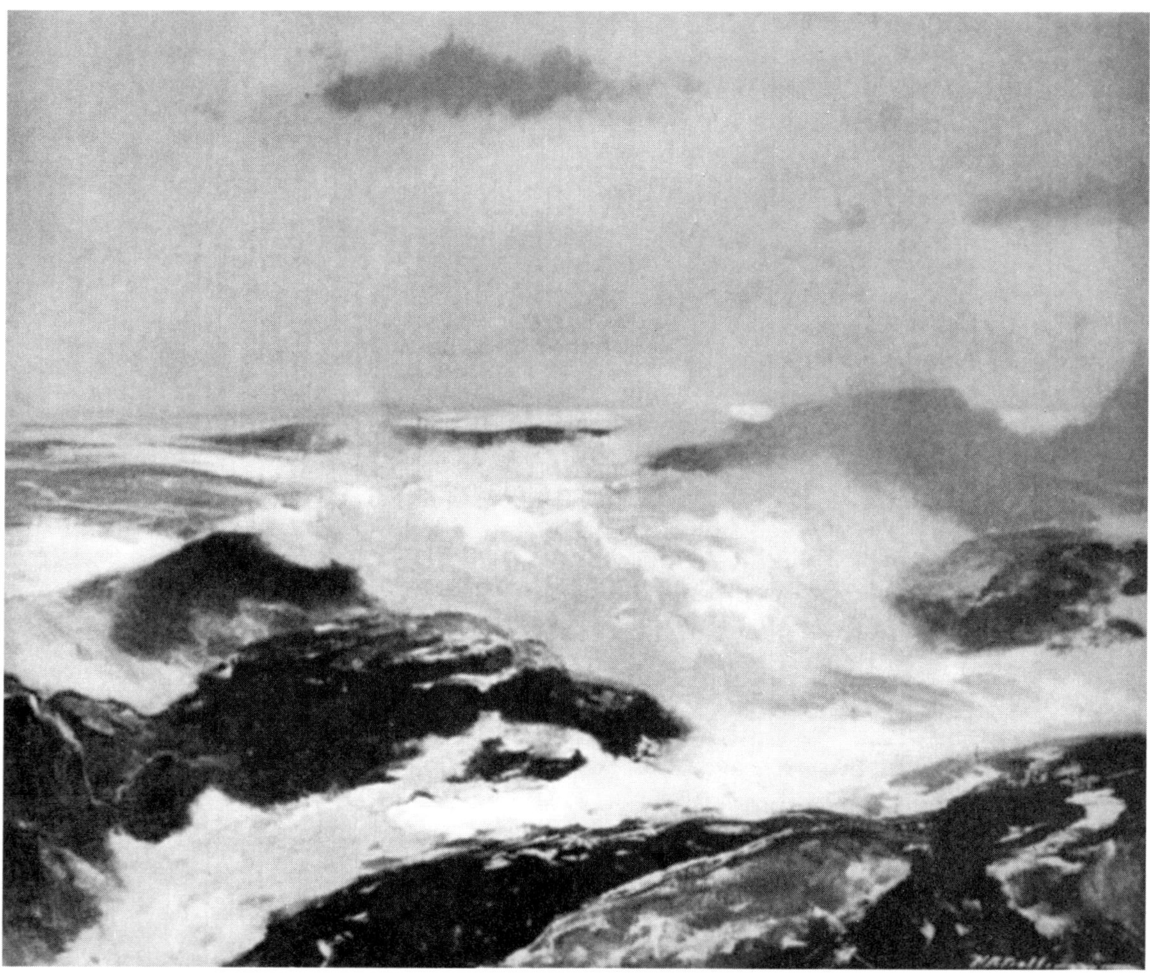

Fig. 43. *Coordination of wave movement with water mass*

As the depth of the water changes, you will find that the spot where the waves crest will change also; so it is a good idea to get the wave action painted as quickly as possible. Decide just what you want the wave to be doing and stick to it. You can't change one part of the surf without changing the rest of it, as the action of each portion of the surf is influenced by, and is part of, the general movement of the whole mass of water (Fig. 43). Never paint a wave as though it were all by itself in the picture. Always think of it in relation to the rest of the water.

Next lay in the rocks, but be careful to paint them in simple planes of light and shade in approximately two values.

I notice that in this chapter I am talking mostly about surf breaking on a rocky shore. Due to the varying depth of the water and the influence that the rocks have on the direction of the incoming waves, you will find that the action of surf on a rocky shore is more complicated than on a sandy beach, where the breakers come straight into shore without being deflected by any rocks. However, you will find that your picture of surf on a sandy beach has to be organized in the same manner and you paint it in the same order.

Starting with the important wave that will be your center of interest, you arrange the flattened out waves moving in toward shore as decoratively as possible with the proper spacing between them. The trailing, floating foam from these waves will give interest and design to the foreground. Then the sand and any incidental details that you find, such as tide pools, driftwood, etc., can be indicated (Fig. 44).

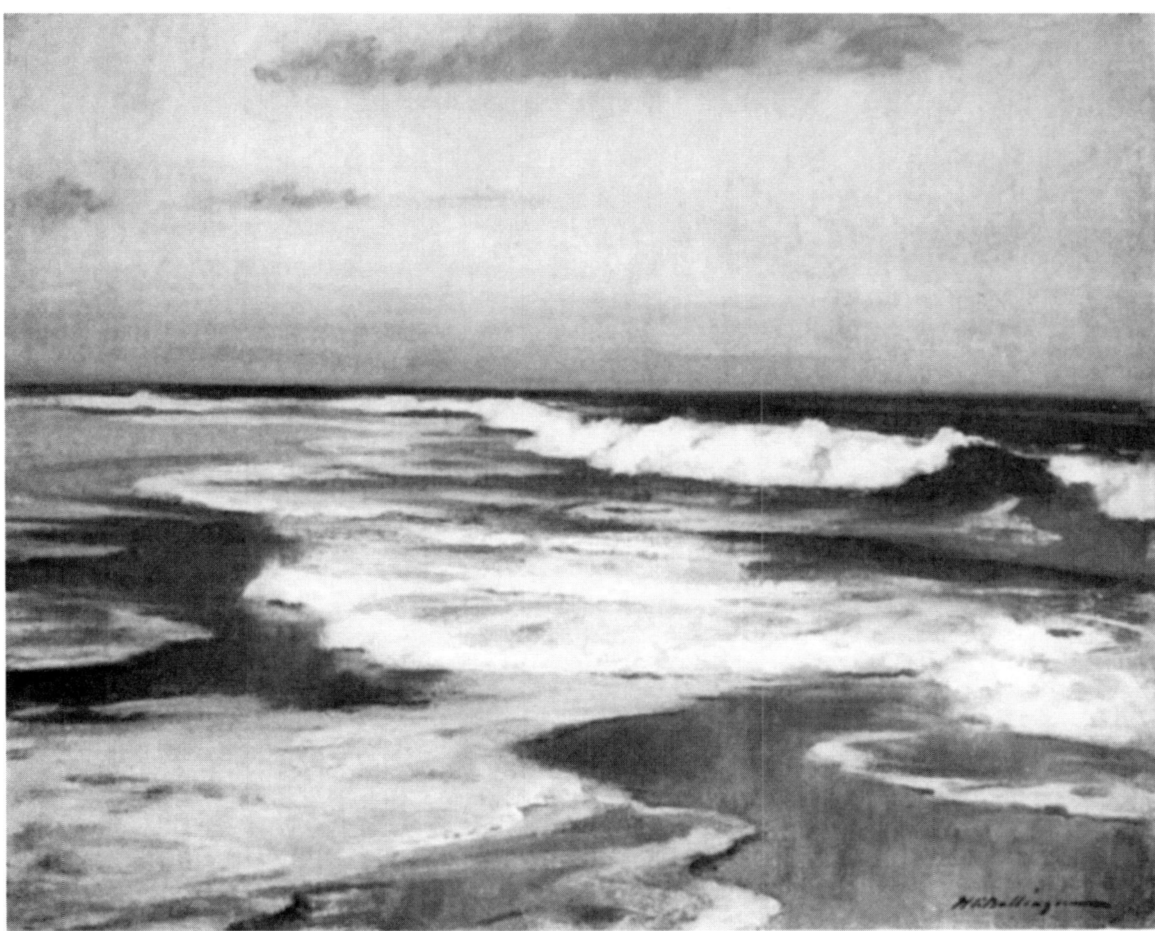

Fig. 44. *Sandy beach with tide pools*

To summarize, when you start to paint, you will start with the sky. Next paint your important wave with the sea around it. Pattern off the flattened out waves with their trailing foam. Then paint the beach and any details that will help give interest to the foreground.

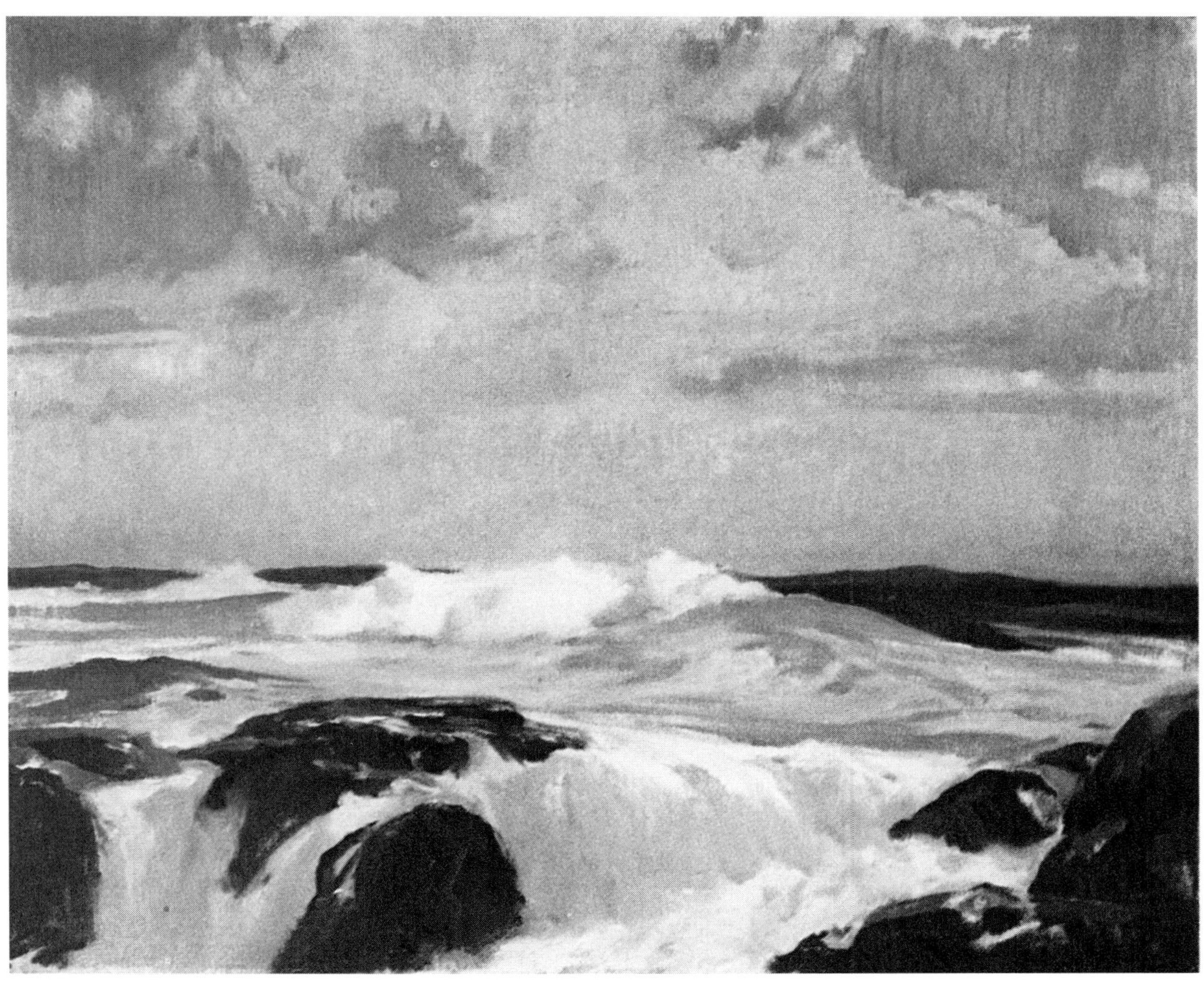

*Cloud pattern in harmony with sea movement, oil, 16 x 20*

(See color insert between pages 42 and 43.)

# 7: Painting Skies

As the sky is the key to the whole picture and should be painted first, I think at this point we ought to talk a little about how to paint it.

A plain blue sky isn't just blue all over, but is a darker, more intense blue overhead. This gradually changes into a paler blue-green as it gets about half way down to the horizon. This in turn fades into more of a pinkish-lavender near the horizon. It is this variation of color and tone which gives the domed look to the sky.

There is a layer of vapor over the surface of the earth which grays-up the sky near the horizon. Over the land this veil is a combination of dust and smoke, while over the sea it is vapor. But in each case it grays-up your color or else occasionally reflects quite brilliant colors from the sun. When you look straight up into the sky you look through the thickness of this vapor layer; consequently you see more of the color of the sky itself.

The warm light from the sun affects the color of the whole sky, so you usually have some pink mixed through the blues instead of just cold blues. When you are looking toward the sun, as you frequently do when painting a picture with back lighting, you will find that the sky is extremely warm and glowing in color near the sun. In fact, the sky could be painted in tones of warm pinks and yellows with hardly any blue showing at all.

Under some conditions the sky directly opposite the sun will receive a lot of warmth from it. When I am painting a picture with a side light I generally warm-up the sky as I get nearer the side from which the light is coming.

Remember that the sky looks darker overhead and gradually lightens in value as it approaches the horizon.

When you are painting clouds, you have a more complicated problem. You have to think of their design, color, and value. A cloudy sky gives you a fine chance to improve the general decorative design of your whole picture. This can be done by arranging the shape of the clouds to carry the interest from the surf to the upper part of the picture. As a sky can add so much to the picture, I believe in showing a generous amount of it. I rarely paint pictures with extremely high horizons. Plenty of sky seems to me to give bigness and atmosphere to the picture.

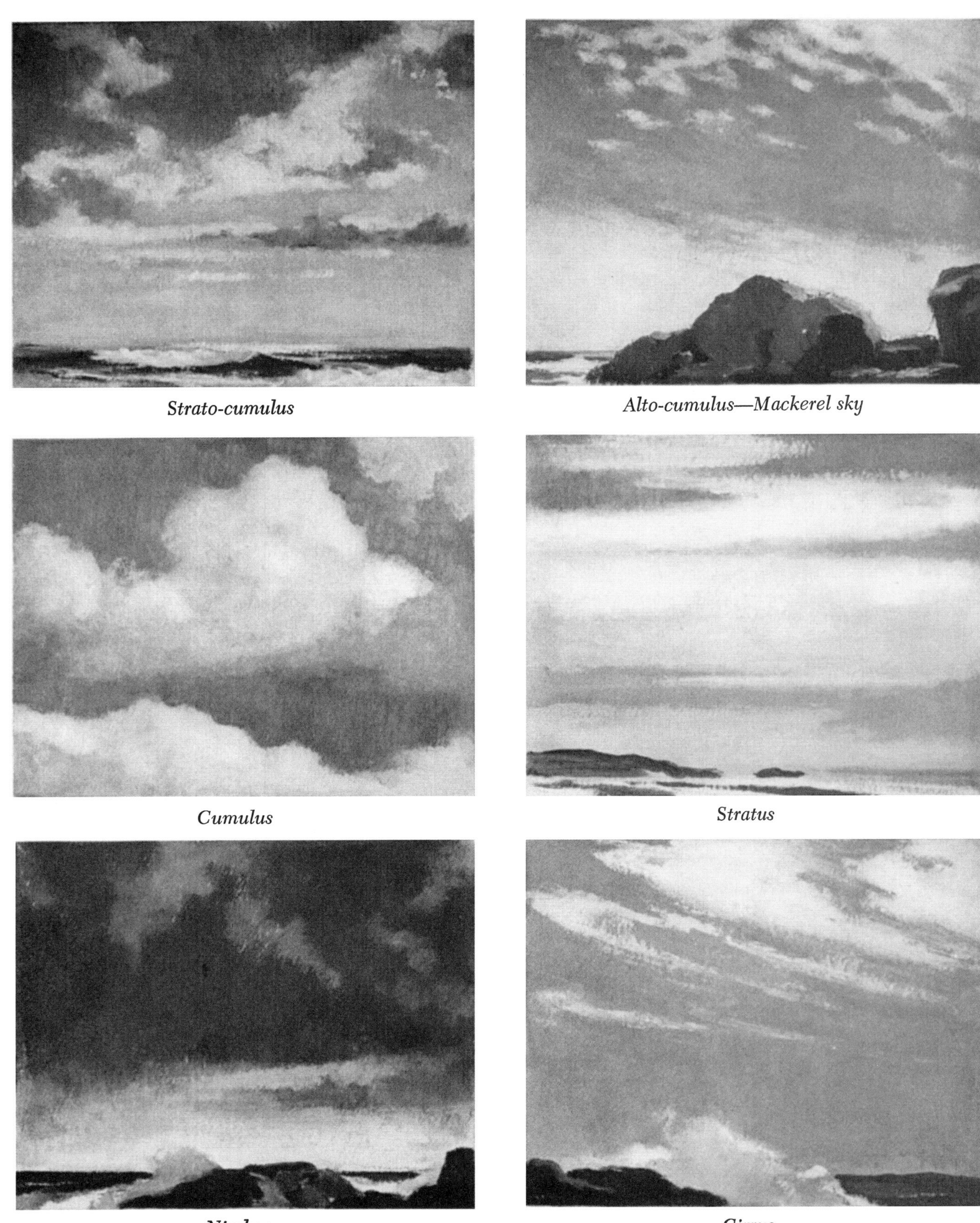

*Strato-cumulus*

*Alto-cumulus—Mackerel sky*

*Cumulus*

*Stratus*

*Nimbus*

*Cirrus*

Fig. 45. *Types of skies*

In order to paint clouds convincingly, it is necessary to study their formations. One may know the correct names for the different types of clouds; however, the important thing is to be able to draw them as they look and paint them in their proper color and value.

Clouds form and float at approximately three different levels; high, middle, and low. They have Latin names which make them sound more impressive. After spending three unhappy years studying first year Latin without much success, I finally decided to let the Latins mess around with their own language. Although I'm not much of a Latin scholar, here, at least, are the names of the main types of clouds:

The low clouds consist of: cumulus—the round wooly fair weather ones; stratus—a layer-like cloud lying quite low; and nimbus—a rain cloud.

You can have clouds that are a combination of two of these three types: cumulo-nimbus or thunderhead; nimbo-stratus; or strato-cumulus, which cover most of the sky in a simple dark mass.

The middle level clouds are called alto. These can be alto-cumulus, which sometimes form a mackerel sky. Alto-stratus are layer-like clouds at a middle level.

The higher clouds are called cirrus and are inclined to be thin and wispy. There are cirro-stratus which are in layers or sheets, and cirro-cumulus which lie in ridges or ripples across the sky (Fig. 45).

Even if you can't remember these fancy names for the clouds, try to remember the characteristic shapes of the important types.

When you are painting a cloudy sky, you will notice that the clouds higher up in the sky, more nearly overhead, are brighter and have more color in them. They can be painted in more sweeping curves, while the clouds as they approach the horizon line become grayer and closer together in more horizontal lines. They, of course, become smaller the farther down they get, near the horizon. Just above the horizon, owing to the previously mentioned layer of vapor, the clouds disappear as individual masses. So there are no clouds just above the horizon; only one fairly light pinkish-blue tone (Fig. 46).

In Chapter 6, I said that the sky should be toned down a little so that it wouldn't detract from the brilliance of the water. This is always the case unless you are painting a sky picture with emphasis on the sky itself. In that case, the surf and sea have to be toned down in order not to detract from the brilliance of the sky. You should have only one main center of interest in your picture.

When you are painting clouds, try to keep many of their edges rather soft and mushy. They have volume, light, and shade; yet they look as if they were made of vapor, not as though they were made of tin. I don't

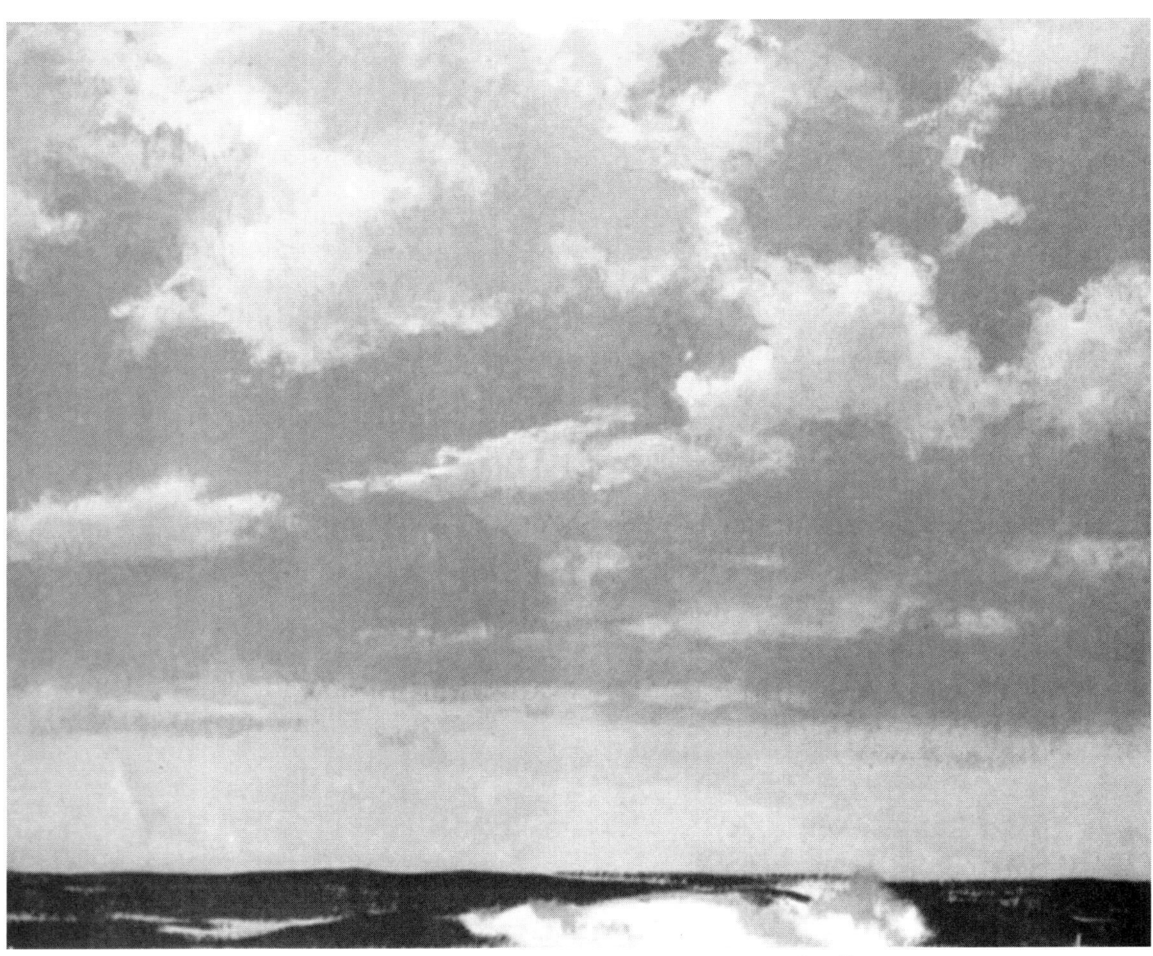

Fig. 46. *Aerial perspective in clouds*

believe that any of us likes a picture in which the clouds appear to be riveted to the sky. Clouds should always have a general decorative pattern that will go with the lines of the rest of the picture. (See color plate facing page 48.)

When I am about to start painting a sky, I carefully study the design of the clouds. When I see an arrangement of their pattern of light and shade that I think would go with the rest of the picture, I quickly sketch it in before it can change.

As you know, clouds, though apparently standing still, are really moving all the time. It is necessary, therefore, if you like a certain effect, to sketch it in as rapidly as you can, because it will change completely in a few minutes.

I like to keep quite a little warmth in the shadows of the clouds, particularly those high overhead. As they get farther away and nearer the horizon, you can use a little less warmth.

Distance cools all colors, as well as graying them up. For example, warm pink rocks would be a lot cooler a quarter of a mile away than if they were in the immediate foreground.

Don't forget while painting the sky, to watch its effect on the rest of the picture. When you have a dark leaden sky, you get a dark sea and black rocks. On the other hand, a bright blue sky produces a blue ocean. The sea is really like a mirror and picks up any changes in the sky.

When painting clouds, use as large a brush as you can, and try to paint as freely and loosely as possible, with lots of soft edges. Paint the sky mostly

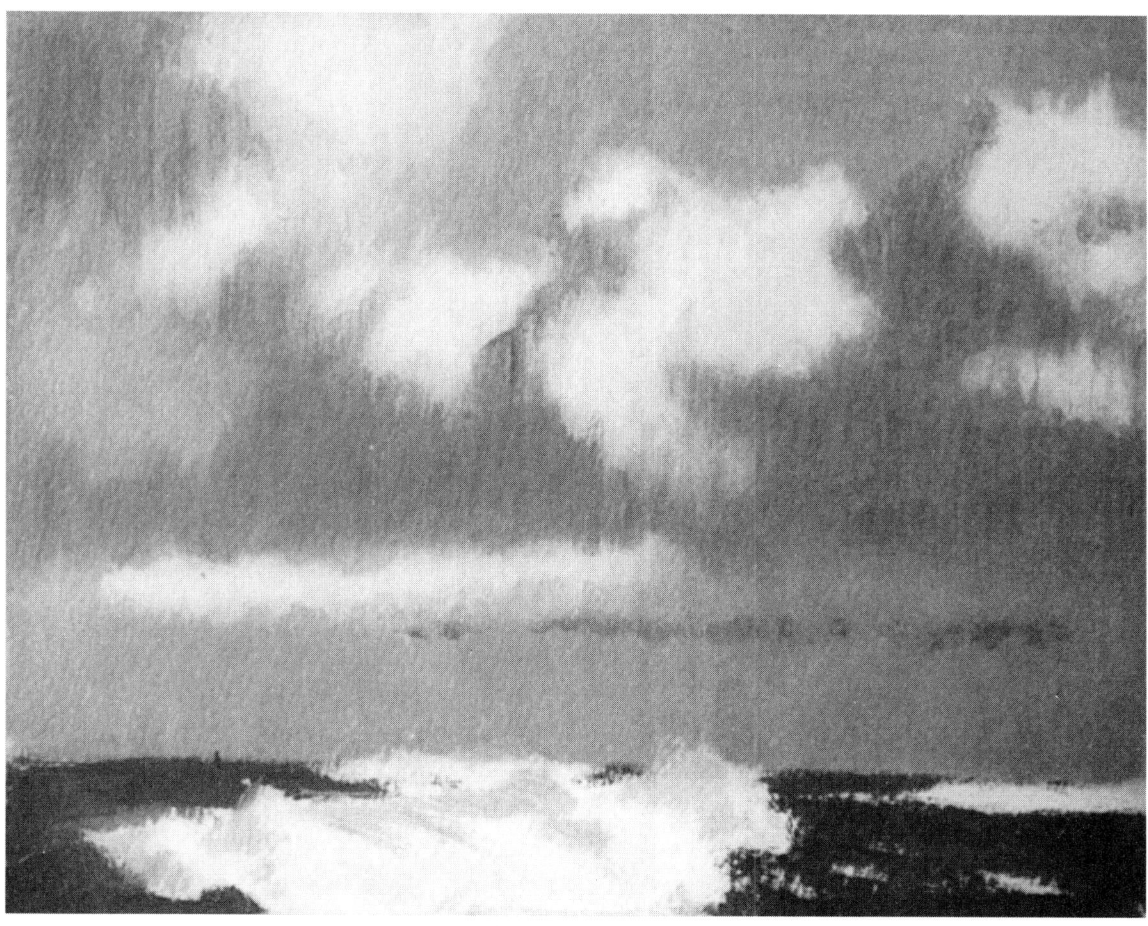

Fig. 47. *Vertical brush strokes in sky*

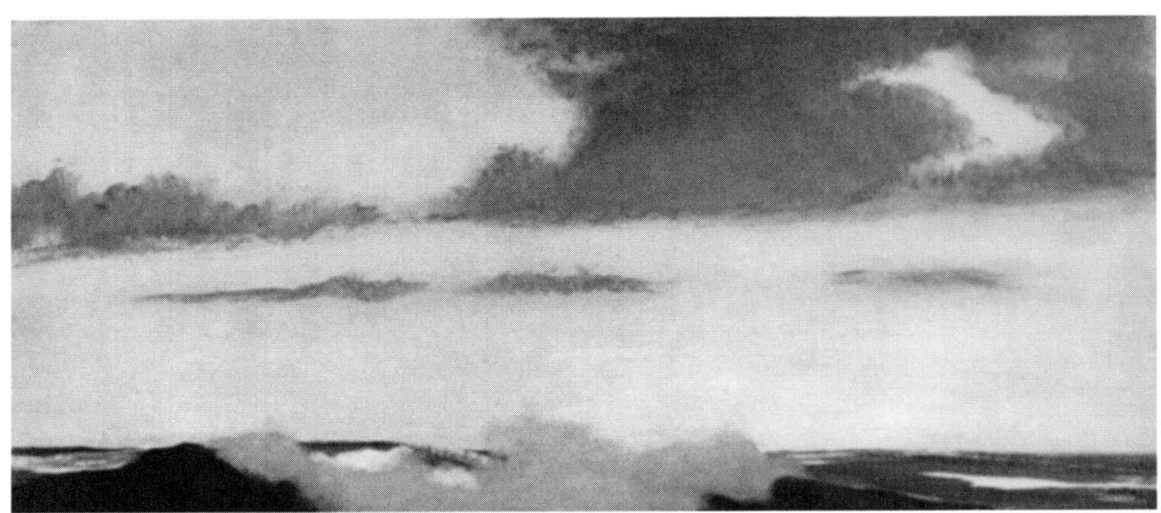

Fig. 48. *Light sky with dark clouds*

with vertical brush strokes instead of horizontal ones. In this way you paint across the action of the horizontal clouds and achieve a looser more artistic effect—it also helps to suggest the domed look of the sky. I try to paint the portions of the clouds in shadow as well as the light portions with the same up and down strokes. Always try to keep shadows as flat and simple as possible with hardly any detail in them; the light areas should be equally flat and simple. You can suggest modeling in the different masses by the degree of sharpness or softness of their edges (Fig. 47).

If you have trouble deciding whether the edge of a cloud is sharp or soft,

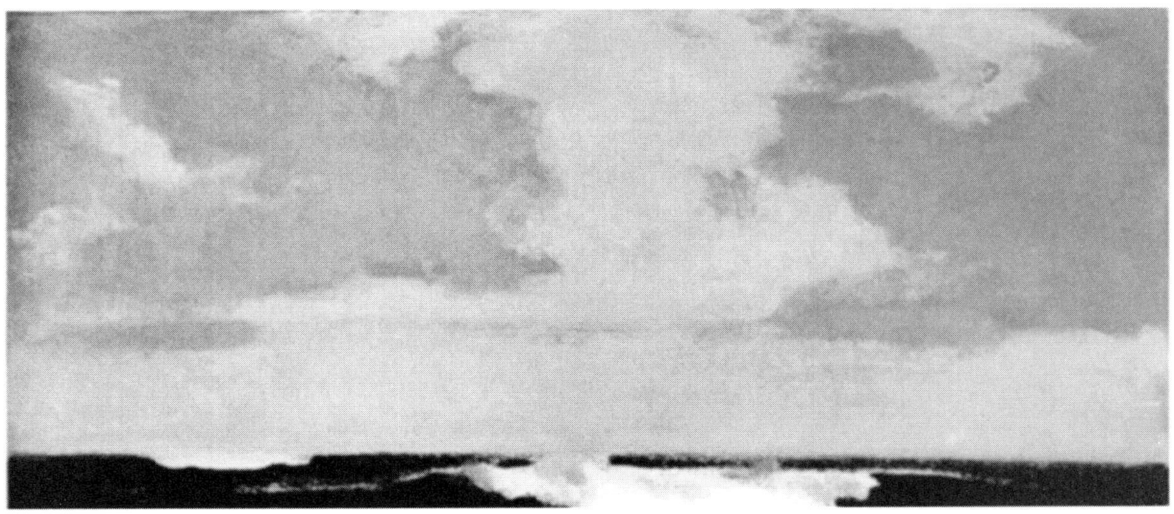

Fig. 49. *Dark sky with light clouds*

just squint at it for a moment with your eyes half closed. Generally the soft edges will melt together when studied in this manner. As a rule, the greater the contrast between two adjoining masses the sharper the edge. With less contrast you get softer edges. If you think that the edge of a cloud is quite sharp, compare it with the edge of a dark rock silhouetted against some foam. You will probably see that the cloud is much softer. In fact, most cloud edges are very soft indeed.

When you are trying to decide how light or how dark to paint parts of the sky, compare them with the strong darks in the foreground or the brilliant light of the foam in your cresting waves. In this way, you can get the proper values easily throughout your picture.

I generally like a light sky that has dark clouds in it or a fairly dark sky with light clouds (Figs. 48 & 49). Either kind has fine decorative possibilities.

# 8: Painting Surf on a Rocky Shore

You have probably noticed that I frequently repeat some of the same ideas in various chapters, although in slightly different language, for I believe that only by repetition can we get these points firmly fixed in our minds. I consider them so important that I am running the risk of boring you by stating some of these principles over and over again.

Before starting to paint waves cresting on a rocky shore, you must know the direction in which they are moving, as well as their action.

Waves move in toward shore from the open sea in roughly parallel lines, and are evenly spaced. As they near the shore they head directly into each bay or inlet of the coastline, as shown in the accompanying diagram (Fig. 50). When the waves reach the rocks, they surge over the smaller ones, but are deflected by the larger ledges and move around them, changing the direction of their forward movement and coming into shore at different angles.

A wave, while having mass and volume, is still fluid and pours around all solid objects in sweeping curves (Fig. 51). This is why it is possible to give

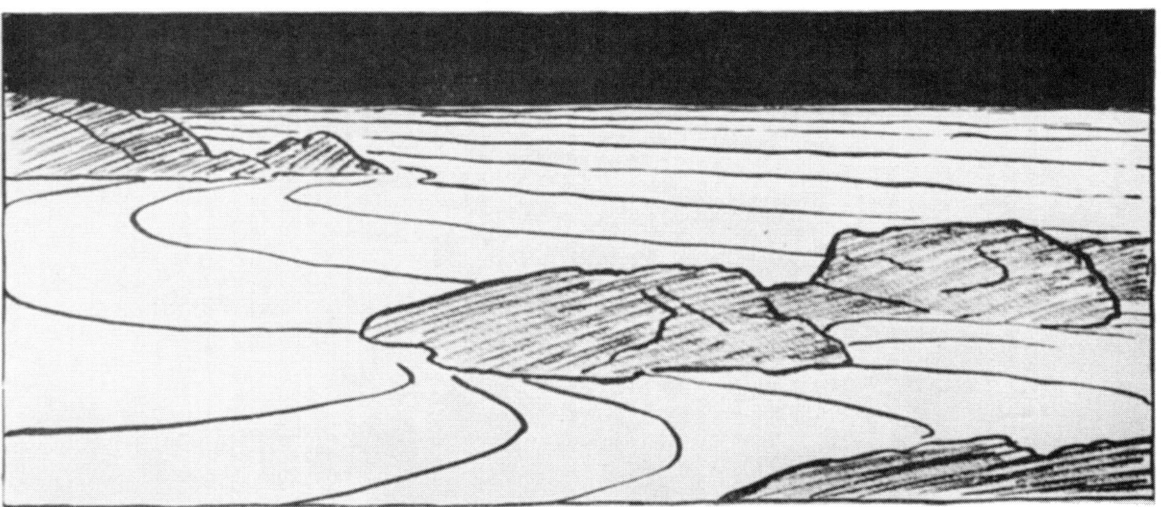

Fig. 50. *Direction of waves coming into shore*

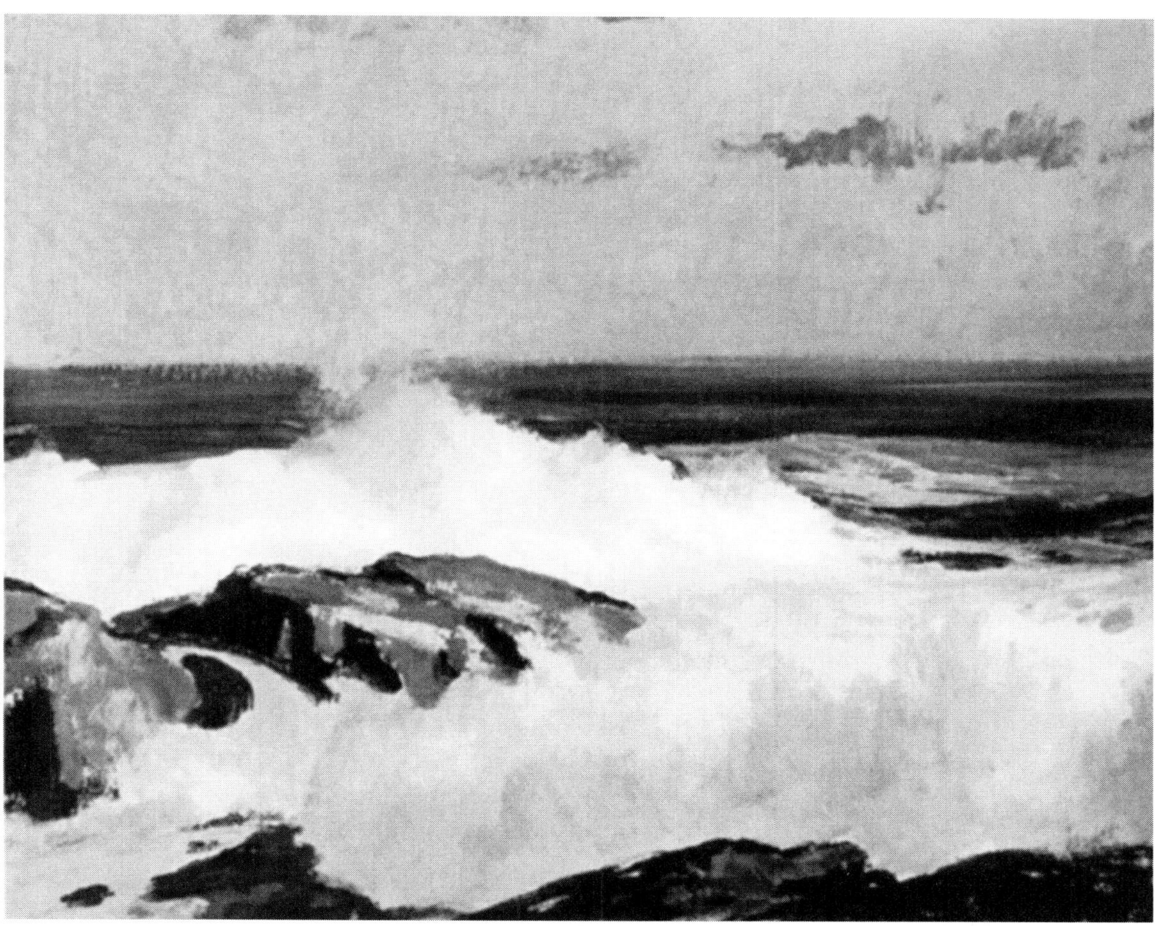

Fig. 51. *Wave swirling around rocks*

a fine, decorative design to the shapes of the breaking waves, and to the floating foam as it swirls around the rocks.

You have to paint the surf from memory, as it changes so fast it is almost impossible to get more than a fleeting glimpse of what the wave is doing. So, as I said above, if you know the direction in which the waves are moving and just what the action is, you will find it easier to make a convincing statement of the scene before you.

*Step One:* When you have decided just what you wish to paint, sketch in outline the shapes of all the important masses of light or dark in the picture. You will then know just exactly what your composition is going to look like.

After studying the surf for a few minutes, arrange the wave action in your sketch to correspond with the real movement of the sea. Plan the entire wave action at once and then stick to this design. Draw in the rocks and the kind of cloud formation that you see in the sky.

*Step Two:* Paint the sky in full value, using care not to make it too bright or important. Your real center of interest should be the water, unless as previously mentioned, you are painting a picture with the emphasis on the sky.

*Step Three:* Paint the breaking wave that is your center of interest, then the rest of the water around it. You can, at the start, suggest the modeling of the cresting wave by painting the shape of the shadows with a light blue-gray color and leaving the white canvas to suggest the light areas.

Be sure to space your waves correctly and remember that wherever you have a wave you will have the trough alongside of it.

*Step Four:* Complete the rocks and the water next to them. Be sure that the rocks look as though they were really in the water and not just sitting on top of it. There will be reflections of the rocks in the clear water in front of them. Sometimes parts of the rocks underwater will be visible through the water if it isn't too deep. (Steps 1-4). (See plate 2 between pp. 42-43.)

Try to paint rocks in simple planes of light and shade in approximately two values.

Wet rocks are darker than dry ones and have more of a glitter to them when the light is reflected from their wet surface.

Rocks are mostly a warm gray color, though they vary in different localities. At least be careful not to make them too hot and eggy-looking in sunlight.

It is a good idea to keep most shadows on rocks a little on the warm side unless they are cooled off by the cold light reflecting down onto them from the zenith of the sky.

Leave the rocks looking a little unfinished. It is better to understate them. If you give them a high finish you can never get the water to look as convincing as the rocks, because the surf won't stand still to be painted.

Shadows in foam are generally a little cool if there is some warmth in the light areas of the foam.

*Step Five:* Finish up the picture. Paint in the light portions of the foam, add detail to cresting waves and floating foam. Also develop any secondary waves in the water. Repeat some of the sky color on the horizontal surfaces of the water. Next pull the sky together, softening edges where necessary.

Add accents either of color or of dark and light contrast to the foreground or your center of interest, and tone down any portion that conflicts with your center of interest. (See plate 3 of color insert between pp. 42-43.)

Then take the picture home and try it in a frame. Generally a frame

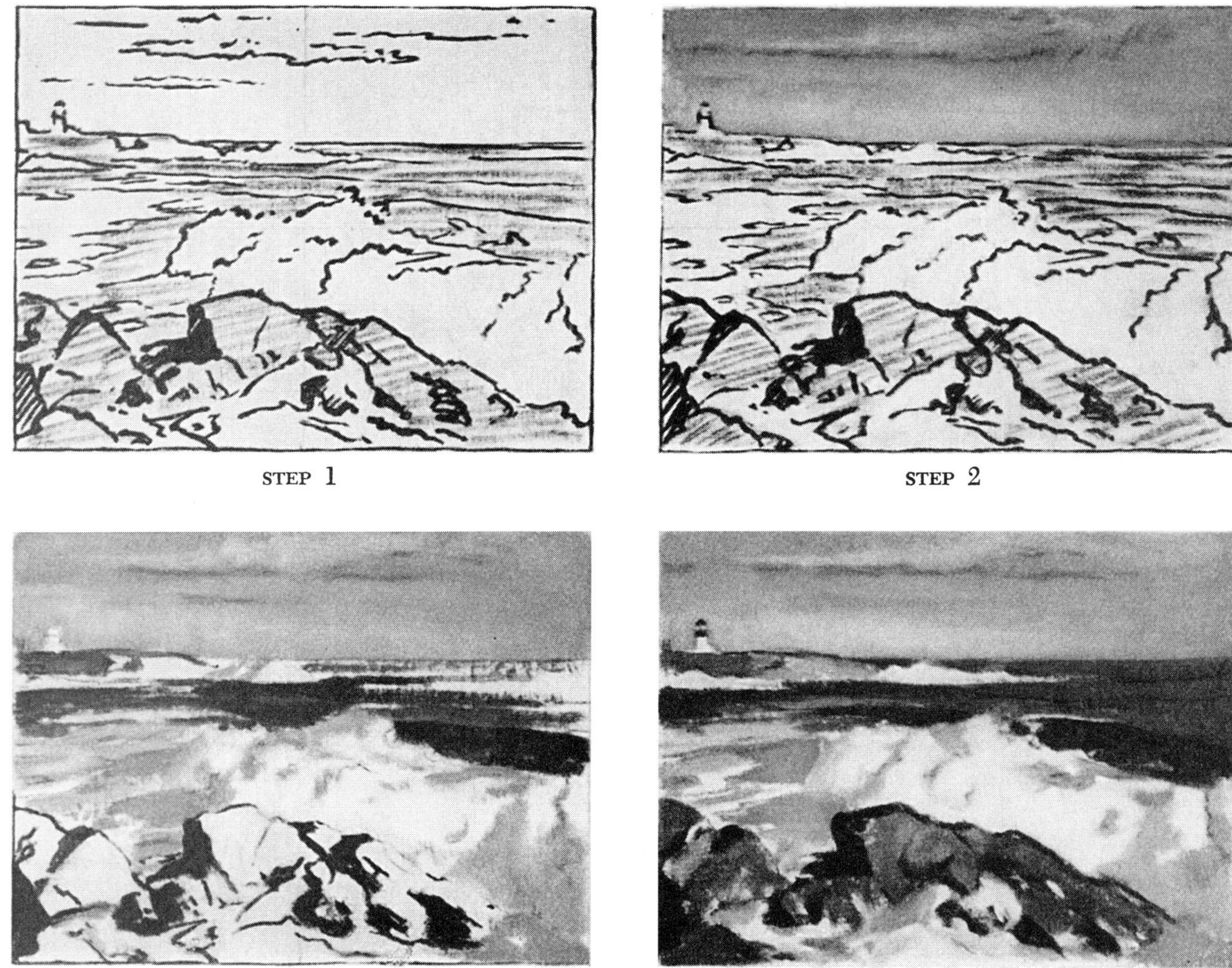

STEP 1

STEP 2

STEP 3

STEP 4

(See color insert between pages 42 and 43.)

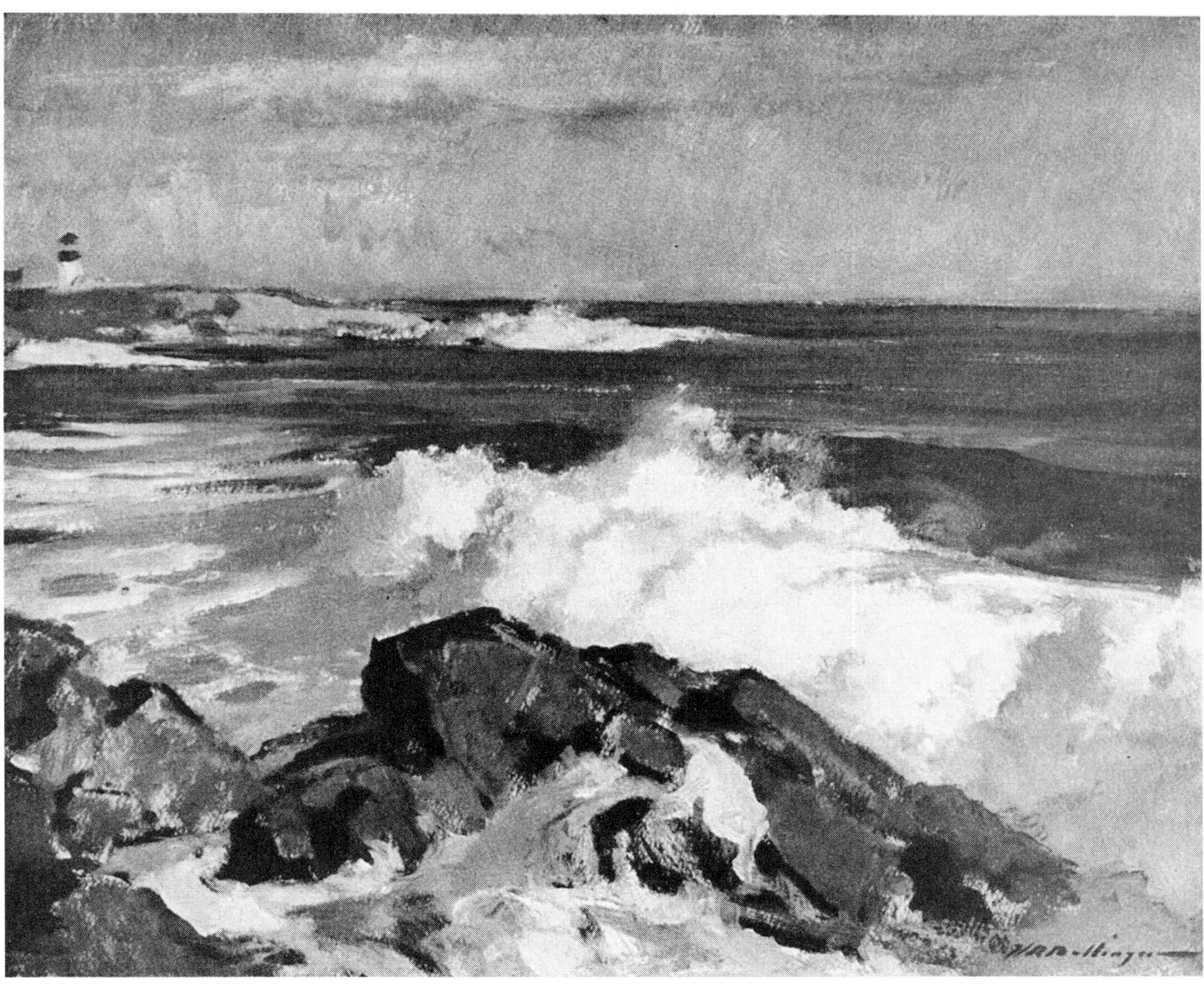

*The completed demonstration, oil, 16 x 20*

(See color insert between pages 42 and 43.)

produces a minor miracle and often makes your picture look much better than you would have thought possible. Occasionally you are even startled to see how good it looks; at other times you are just merely startled.

However, don't be discouraged if your first few seascapes are a little less than the masterpieces you had hoped to paint. It requires a lot of practice and study of the sea to capture it on canvas. I doubt if even Winslow Homer was able to paint his great marine paintings without first messing up quite a lot of practice canvases.

# 9: Painting Surf on a Sandy Beach

In the preceding chapter we were talking about painting surf on a rocky shore. This was really easy, as the rocky foreground was a great help in composing the picture. The rocks with their strong darks and decorative shapes gave you an interesting pattern of darks that contrasted pleasantly with the lighter masses of water and foam swirling around them.

However, the action of the water itself is the most important thing in a marine painting. That is why surf coming in on a flat sandy beach can be just as thrilling as any other seascape. It just happens that most of my own painting has been done along the northern New England coast of the United States, which is predominantly rock bound. But I also enjoy painting in Florida and in the Caribbean, where there are long stretches of sandy beaches.

Waves coming in on a sandy beach, because of the uniform depths of the water, generally crest for a much greater distance, and the wave appears much longer than on a rocky shore with its varying depths of water (Fig. 52).

Due to the flatness of the beach, a wave after cresting will be carried

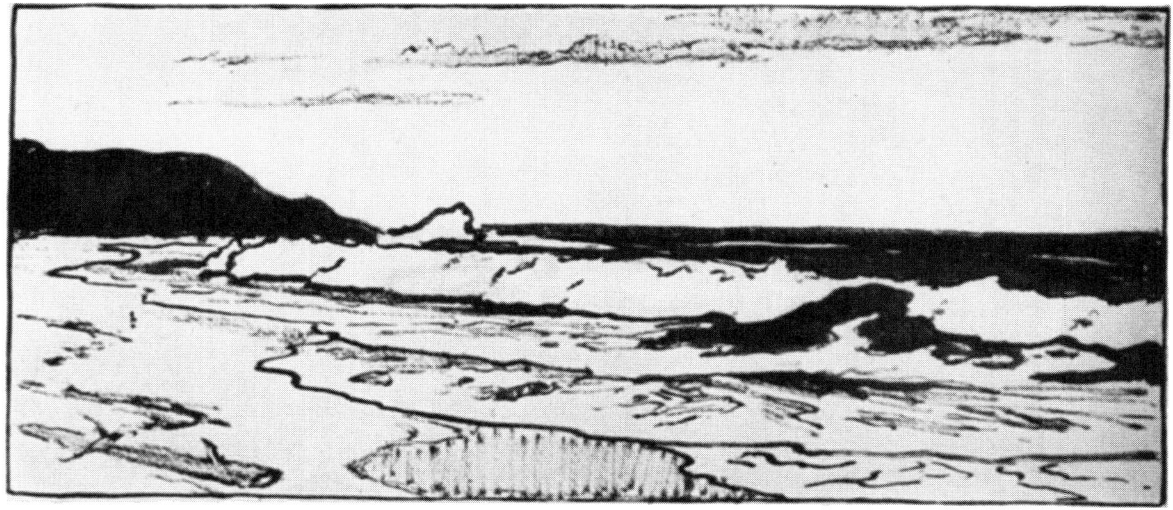

Fig. 52. *Length of cresting wave on sandy beach*

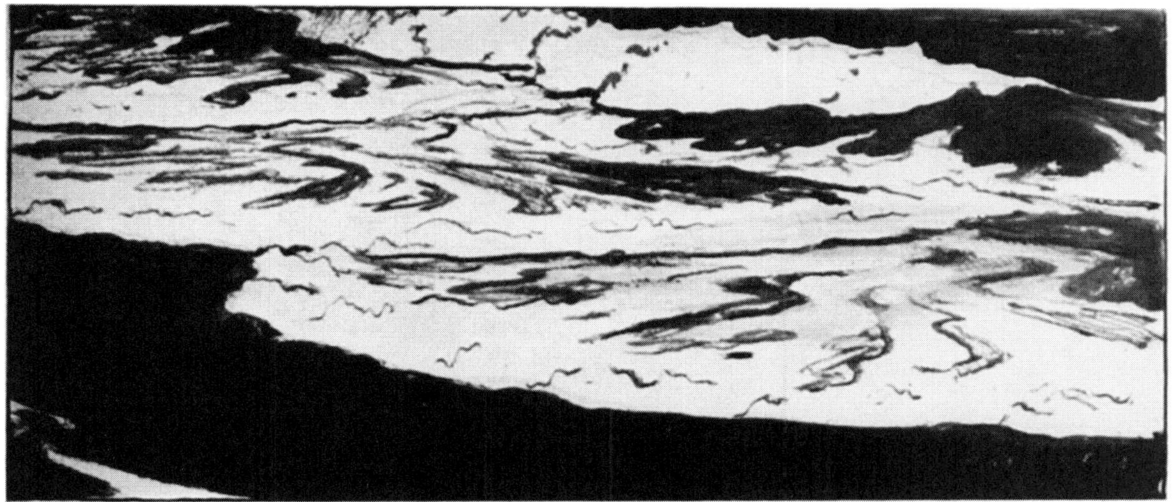

Fig. 53. *Wake of flattened-out wave*

by its own momentum a considerable distance up the beach. It gradually flattens out and leaves a wake of foam trailing out behind it (Fig. 53). When it reaches the end of its forward movement, it turns into floating foam which gradually disappears as the water flows out to sea again.

When the flattened-out wave gets into shallow water, it frequently changes its direction and continues inshore at a slightly different angle. This gives the effect of one wave, with its layer of water flowing over the wake of another wave at irregular intervals and at different levels. This produces a layer-like formation (Fig. 54).

As long as a wave is still moving forward toward the beach, there is height and thickness in its front edge, while the trailing foam behind it is flatter and more lacy. The flattened-out curves of the front edges of these waves will help give interest and pattern to the foreground (Fig. 55).

The sand at the water's edge is darker when it is still wet from a preceding wave. This dark note will contrast with the light foam of the nearest wave, and will add weight and interest to the foreground.

Sometimes interesting lines are left on the sand from preceding waves. These lines are roughly parallel to the edge of the water, and are made by seaweed, driftwood, and other debris. Tidal pools in the sand or even small creeks trailing out toward the sea give variety and interest to the foreground.

As I have said earlier, you will have more interesting lines in your picture if you look along the beach and paint the scene at an angle, rather than looking straight out to sea.

Because there isn't as much interesting detail on a sandy beach as there is on a rocky shore, you have to play up the design of your flattened foreground waves with their trailing pattern of floating foam. Look for a fine lacy pattern of the lighter foam against the darker color of the clear water.

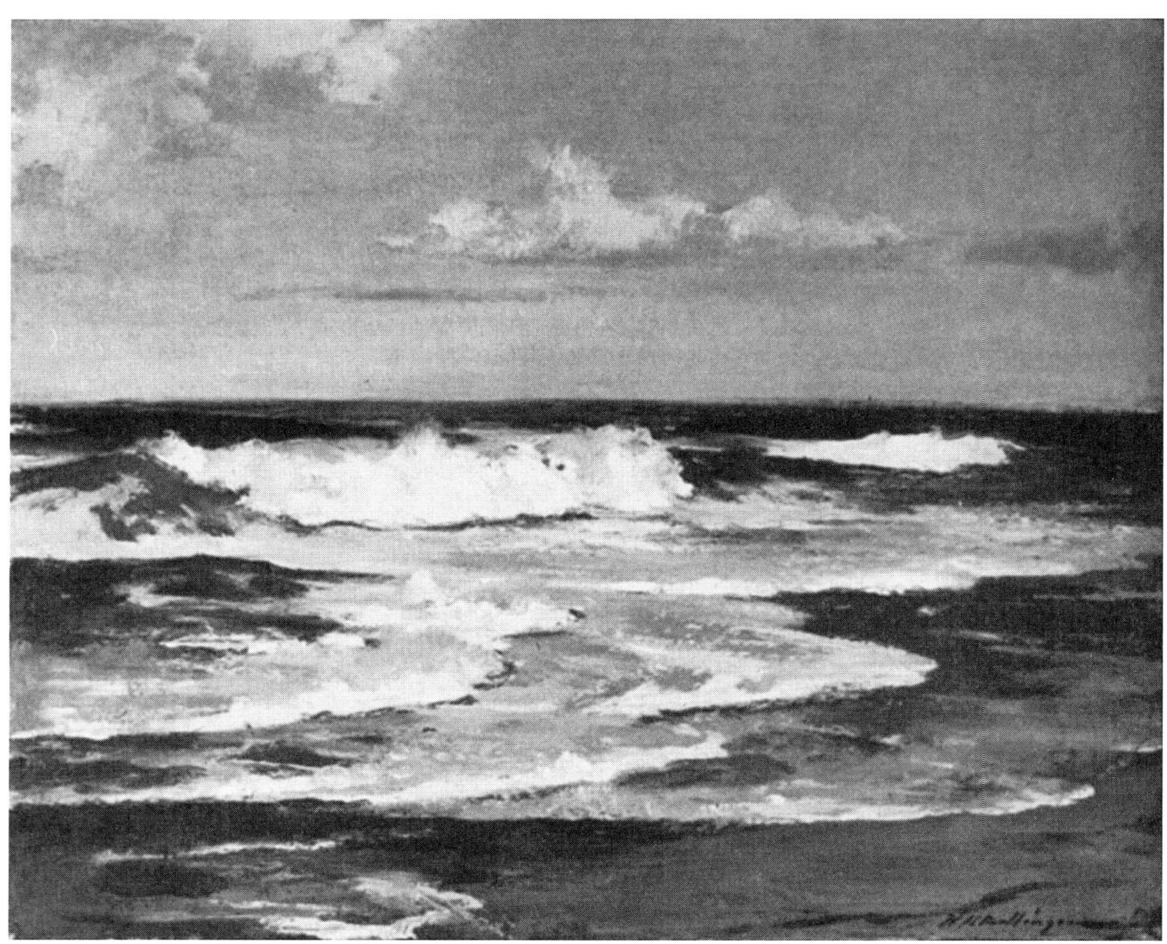

Fig. 54. *Directions of waves in shallow water*

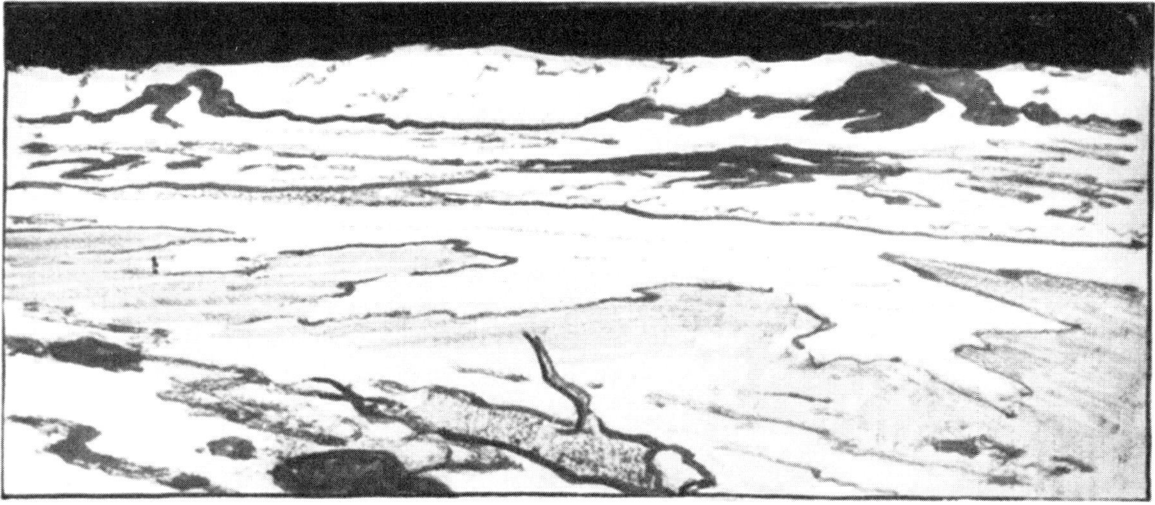

Fig. 55. *Foreground interest*

This will add foreground interest and will lead your eye into the picture.

The shallow water close to shore takes on the warm tone of the sand beneath it, and will be more of a tan color. Farther out, as the water gets deeper and you can't see much of the bottom, the water is more of a blue-green. The clear water of your cresting wave will be a rich, dark blue-green as you look straight into it and can see the color of the water itself.

The sky is extremely important in a sandy beach picture, and should be made as interesting as possible. Whenever you have clouds in a sky, you can have cloud shadows over parts of your picture if you so desire. I think cloud shadows are God's gift to the artist, as they enable him to concentrate his light exactly where he wants it; also to have dark shadowy spots anywhere he needs darks, either on the land or on the water.

*Step One:* Decide on the black-and-white pattern of the picture. Sketch in outline the shapes of all the masses of light and shade, from the sky down to the foreground.

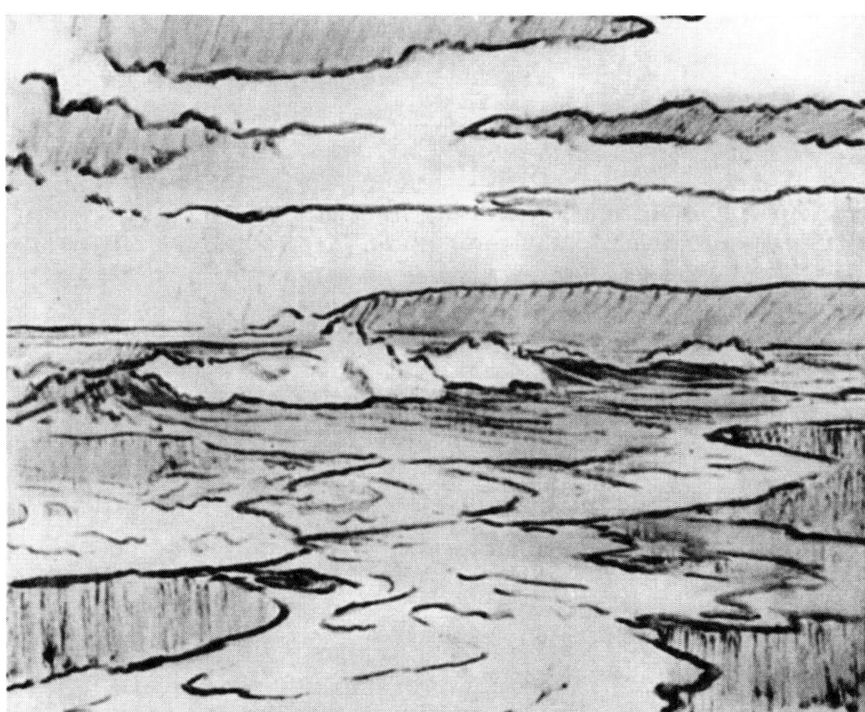

STEP 1

*Step Two:* Paint in the sky, being careful to get the right value and color.

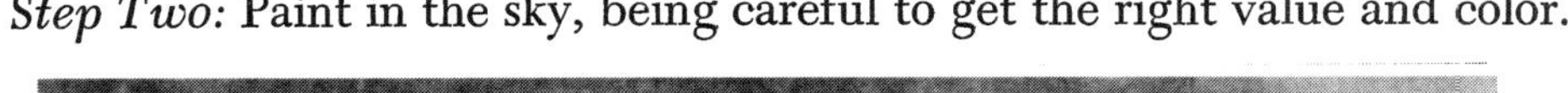

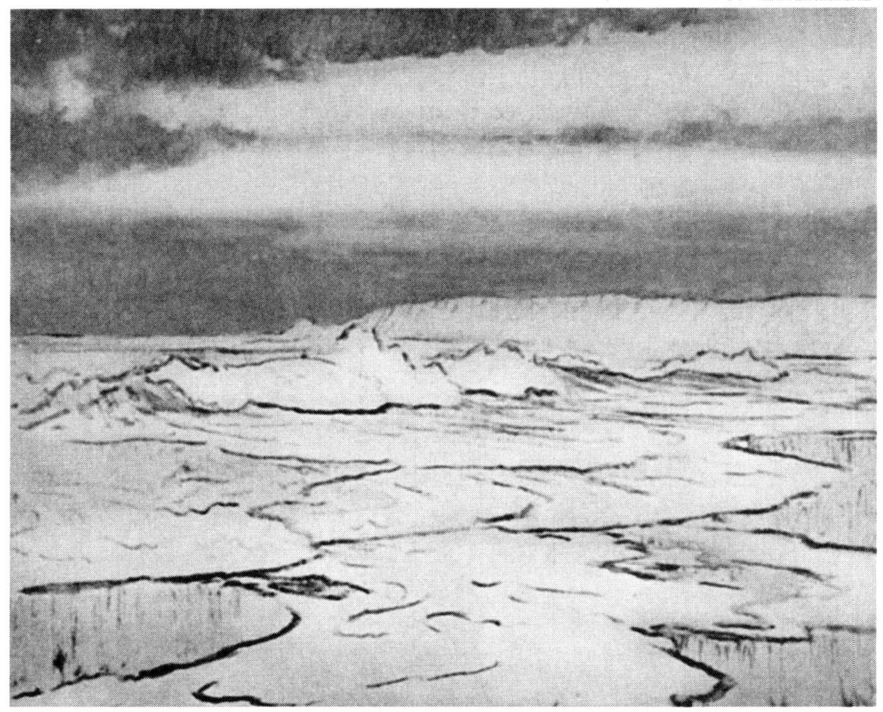

STEP 2

*Step Three:* Paint in the big, cresting wave of your center of interest; also the flattened-out waves in the foreground. Try to get the shape of your floating foam at the same time, as well as the pattern of the surrounding water.

*Step Four:* Paint the beach. Try to improve the pattern of your cresting waves and floating foam, as well as the distant sea (Steps 1-4).

*Step Five:* Finish the picture. Add accents of color or tone to the center of interest and to the foreground. Pull the whole picture together by toning down unnecessary light spots and losing details wherever possible. (See plate 4 of color insert between p. 42-43.)

A good picture should have a fine, decorative pattern of light and shade. Equally important is a harmonious and pleasing color scheme.

I like a picture with comparatively few colors. I think it is generally stronger and more harmonious than one with a great many contrasting

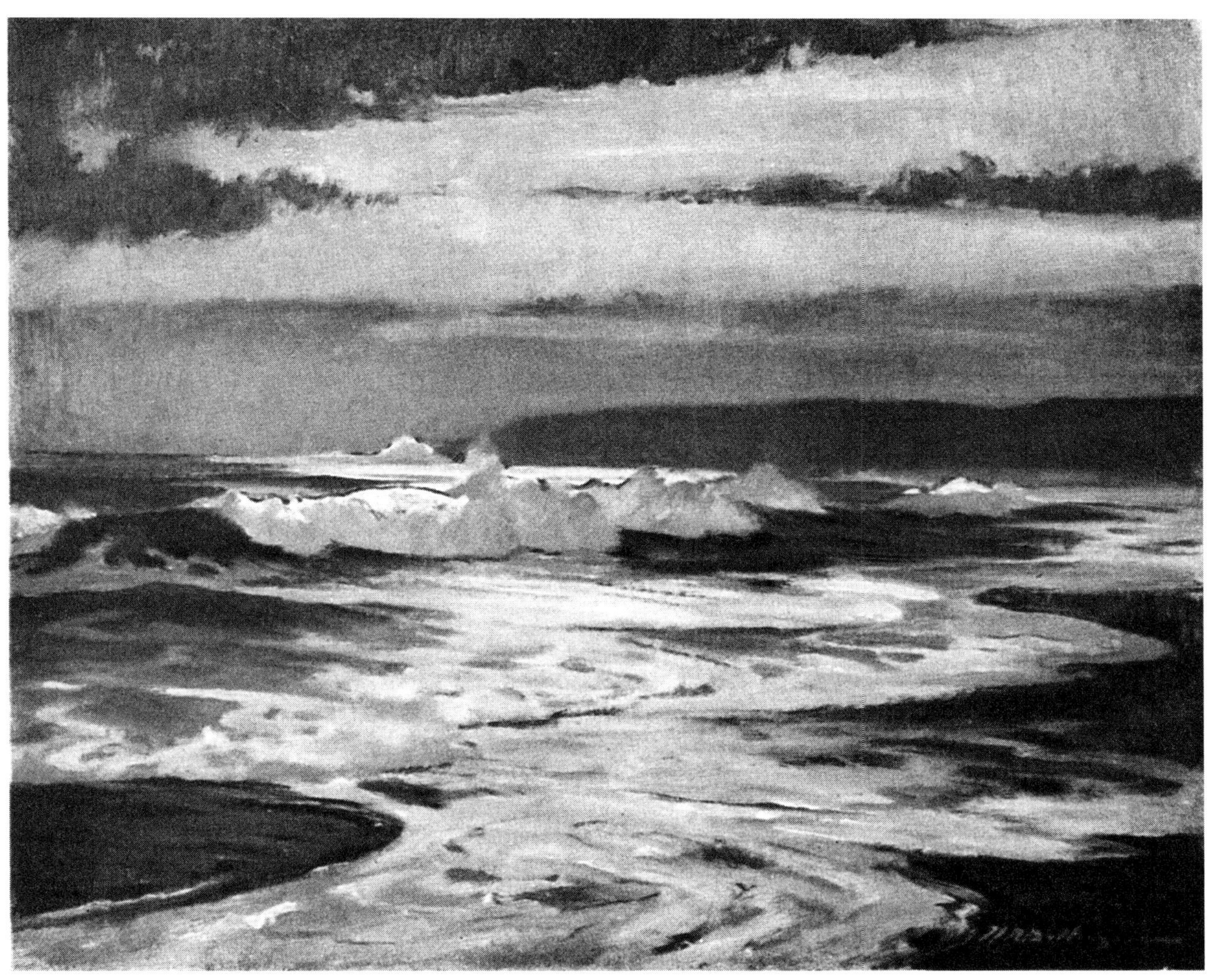

*Demonstration of painting running surf on a sandy beach, oil, 16 x 20*

(See color insert between pages 42 and 43.)

STEP 3

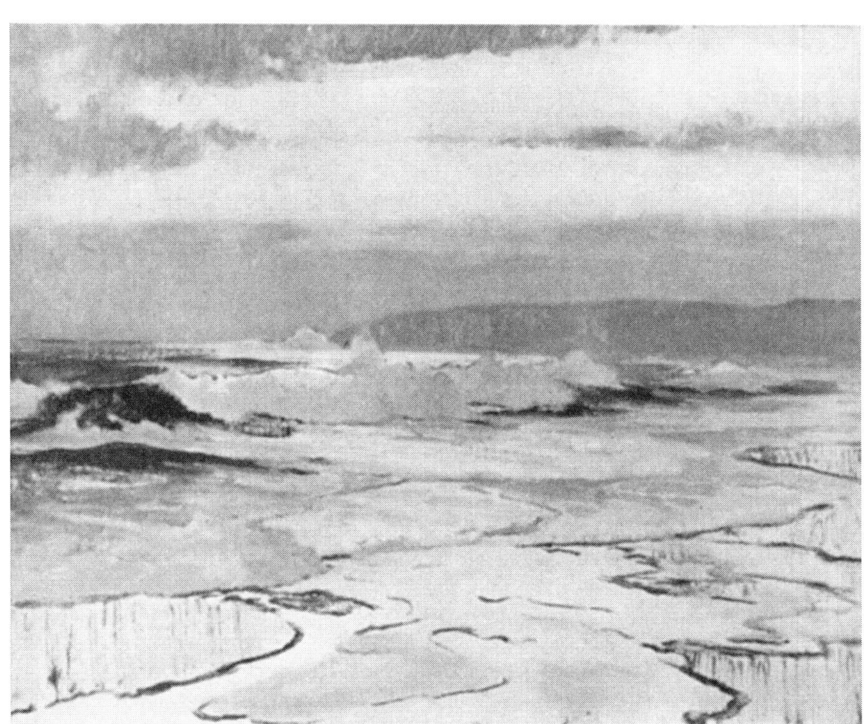

STEP 4

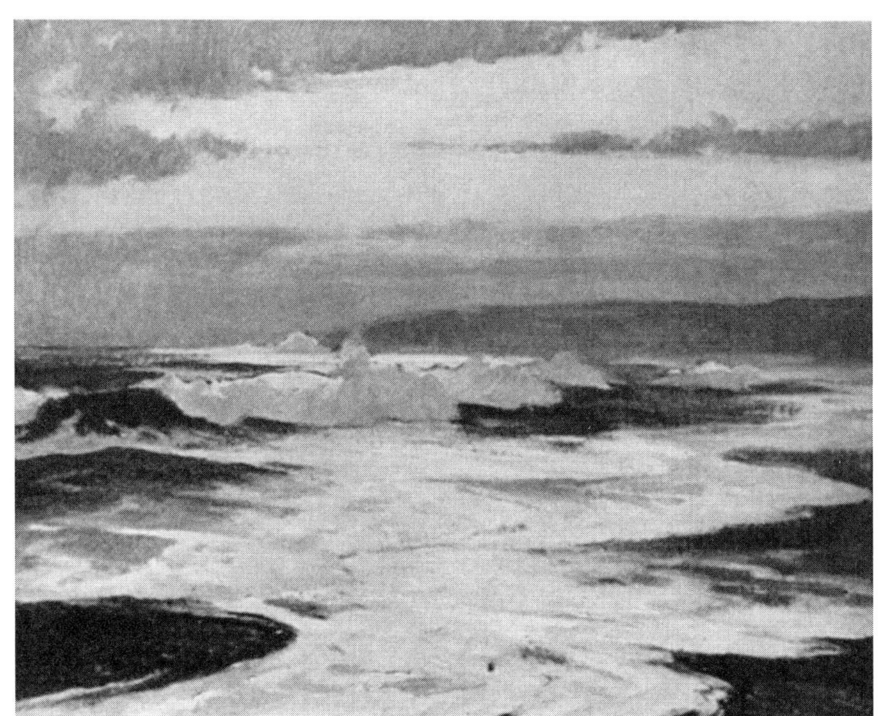

colors. That is why a picture with a back lighting that gives the effect of a warm glow over the whole picture is immensely effective.

Some of the most striking pictures that I have done have been painted almost in black-and-white tones, with a single splash of brilliant color as an accent.

Color is, after all, a very personal thing. No two persons would see exactly the same colors in any given scene. I think that if you just use good taste in your choice of color and avoid raw clashing colors, you should be able to paint some pleasing and harmonious pictures.

# 10: Open Sea Painting

An open sea painting without any foreground interest of rocks or sand has to have a careful arrangement of waves and foam in the foreground to lead your eye into the picture.

When painting the open sea, I think of the big swells as a series of mountain ranges with the smaller, secondary waves leading up to them like foothills in a landscape (Fig. 56). Between each wave or mountain you have a corresponding trough or valley, so space your waves and be sure to get the proper contrast between the height of the waves and the trough between them.

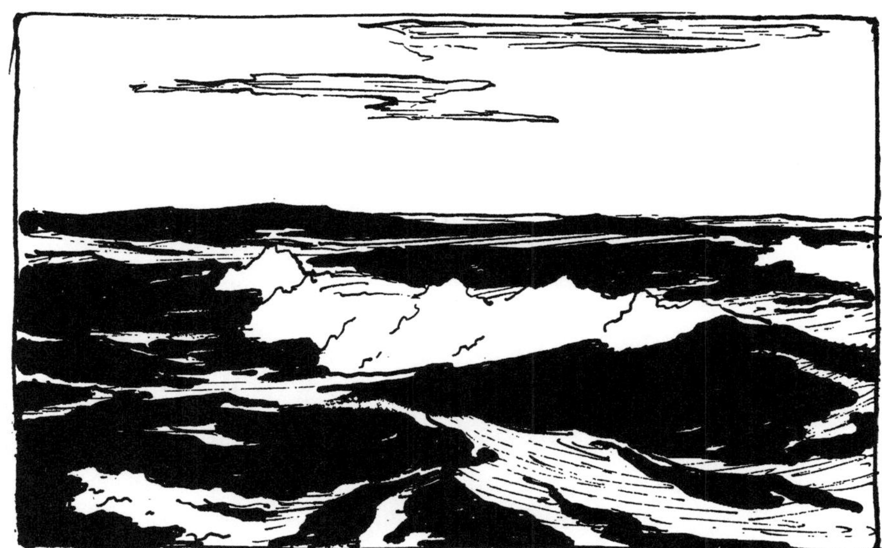

Fig. 56. *Wave arrangement in open sea painting*

A big wave cresting near the center of your picture would provide the usual center of interest. By working out a simple design for your masses of light foam and the large, dark shapes of the uncrested waves, you can work out a fine strong design of lights and darks. The sky becomes increas-

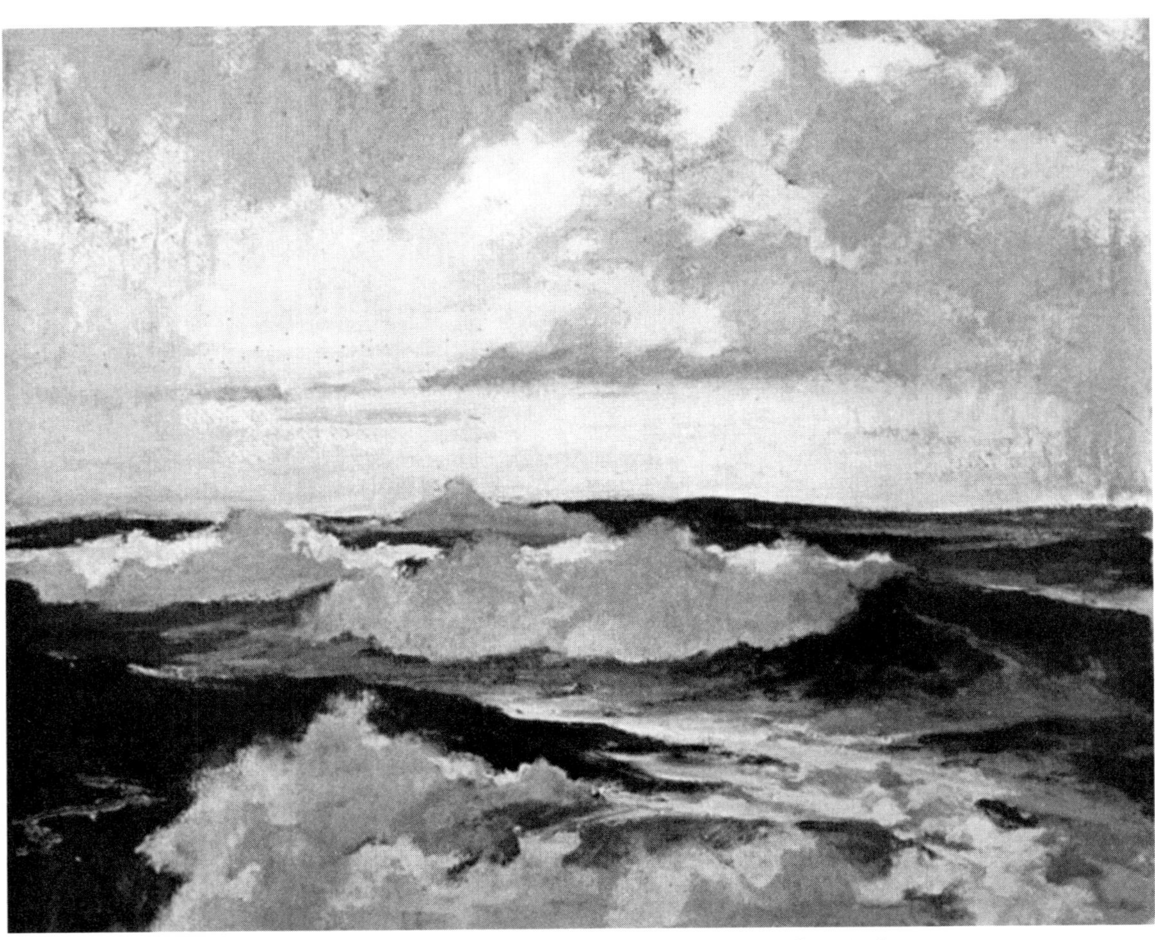

Fig. 57. *Open sea painting with a lot of sky*

ingly important and should have as much interest as possible (Fig. 57).

In most of the open sea pictures that I paint there is generally a boat, a raft, or some human interest. But if you do a good enough job painting the sky and sea, you won't have to depend on boats, seagulls, or people to add interest to your picture. To this end I have known artists even to put mermaids or bathing beauties in their pictures. I think they are always fascinating to look at, but I suspect that a well-painted seascape will have a universal appeal without any added human interest.

The lighting of an open seascape is extremely important and should be completely thought out in advance. A back lighting, you remember, is one in which you look into the light and all upright objects are silhouetted against the light beyond. In this way, you can have sharp, warm reflections of light on horizontal planes, while all the upright parts of waves and foam are in shadow. By having your strongest contrast of light and shade in the foreground and in your center of interest, you will be able to concentrate your light just where you need it.

When you look into the sun, the sky takes on a lot of warmth and you can get a glow of warmth over the whole picture.

A diffused down light is another form of lighting that is effective, as it enables you to paint everything in the picture in two planes, a light top plane and a darker side plane. In this lighting, you get more of the local color of the sky, sea, and waves than when painting with a back light. This is the effect that you get on cloudy, foggy, or rainy days, as I have mentioned earlier.

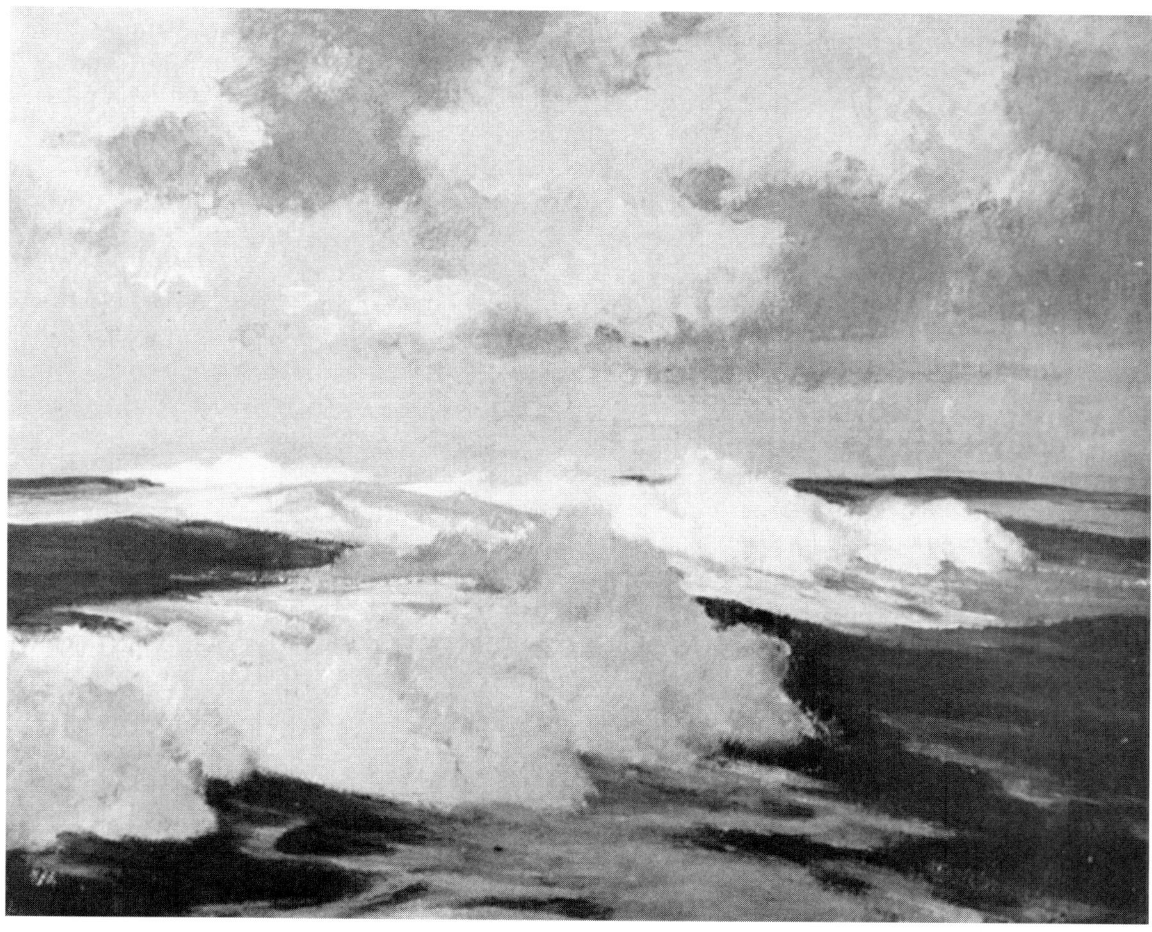

Fig. 58. *Cloud shadows on sea*

Another type of lighting—the side light—is most frequently used. It is always effective, particularly when you combine it with a blue sky that has broken clouds in it. This gives you the opportunity of using cloud shadows in your picture.

As previously mentioned, when you have cloud shadows on the sea, you have a beautiful opportunity to use any portion of the picture as part either of your light pattern or of the dark. A dark wave in the foreground silhouetted against a lighter wave in full sunlight farther out in the picture is always effective (Fig. 58).

*Step One:* Decide on the black-and-white pattern of your picture; outline the shapes of all the masses of light and dark in your picture. An open sea picture is generally painted in the studio from memory, or from quick sketches made at sea. You can, therefore, use any color scheme and light effect that appeals to you. As previously noted, you have to arrange the foreground waves in an effective and decorative pattern of light and shade in order to lend interest to the foreground and to lead your eye back into the picture to your center of interest.

*Step Two:* Paint the sky in the usual way.

*Step Three:* Paint the cresting wave that is your center of interest; then the foreground waves that lead the eye into the picture and the water around them.

*Step Four:* Paint the distant ocean, taking care to see that the color of the sky is reflected in the distant sea. Add detail to the cresting wave that is the center of interest, and further develop the foreground waves (Steps 1-4).

*Step Five:* Complete the picture in the usual way. It is a good idea, when finishing the picture, to stop before it is entirely done. We all have a tendency to overwork a picture. All the thousand and one fancy touches that we add at the last minute seldom improve the picture.

There is an old saying, that it takes two people to paint a masterpiece: one to do the painting, the other one to hit him over the head with a hammer at the proper moment before the artist has a chance to spoil the picture.

I'll bet that our wives (or husbands) would be only too willing to do a little work with the hammer when they see a fine vigorous sketch being ruined by too many finishing touches.

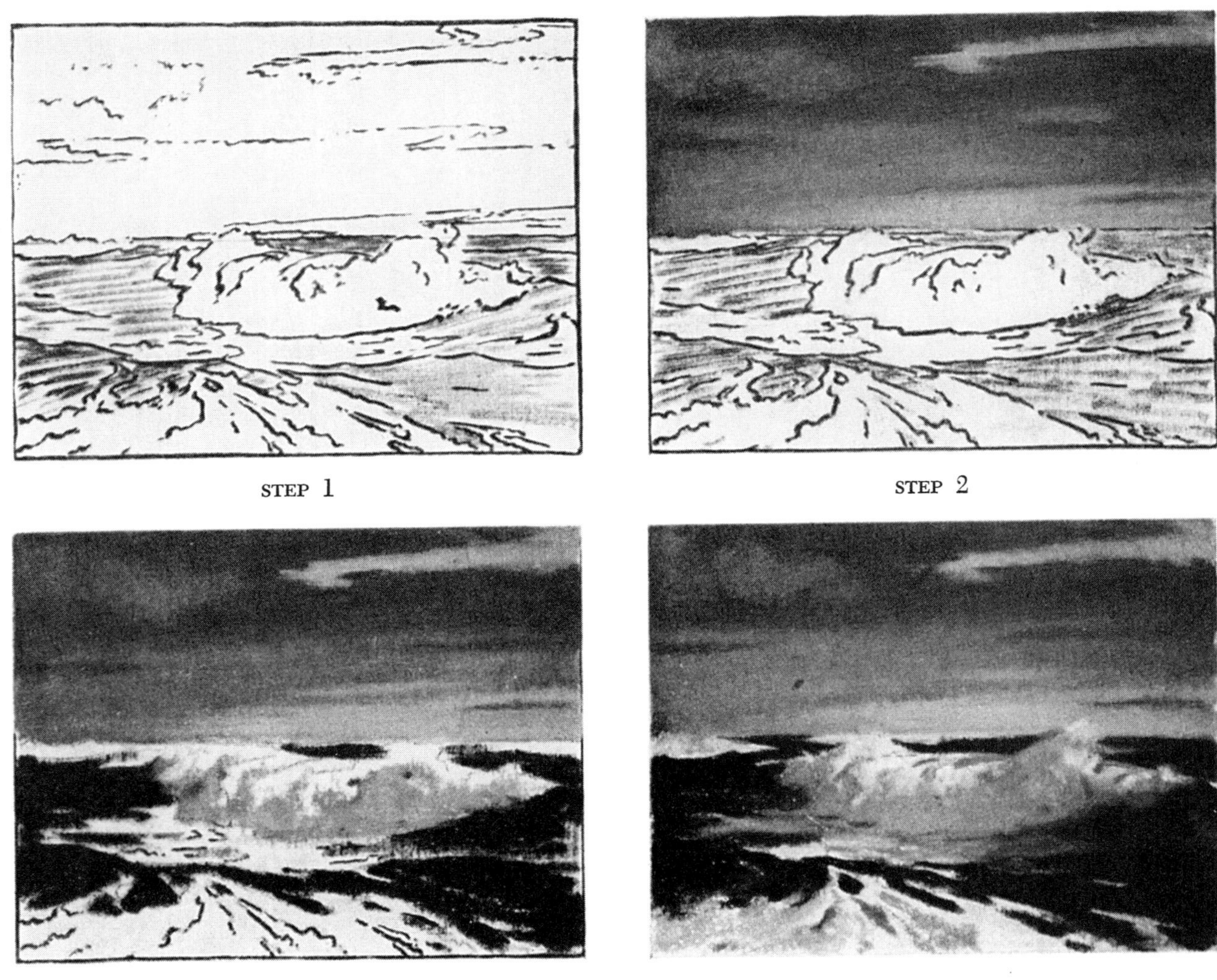

STEP 1

STEP 2

STEP 3

STEP 4

(See color insert between pages 42 and 43.)

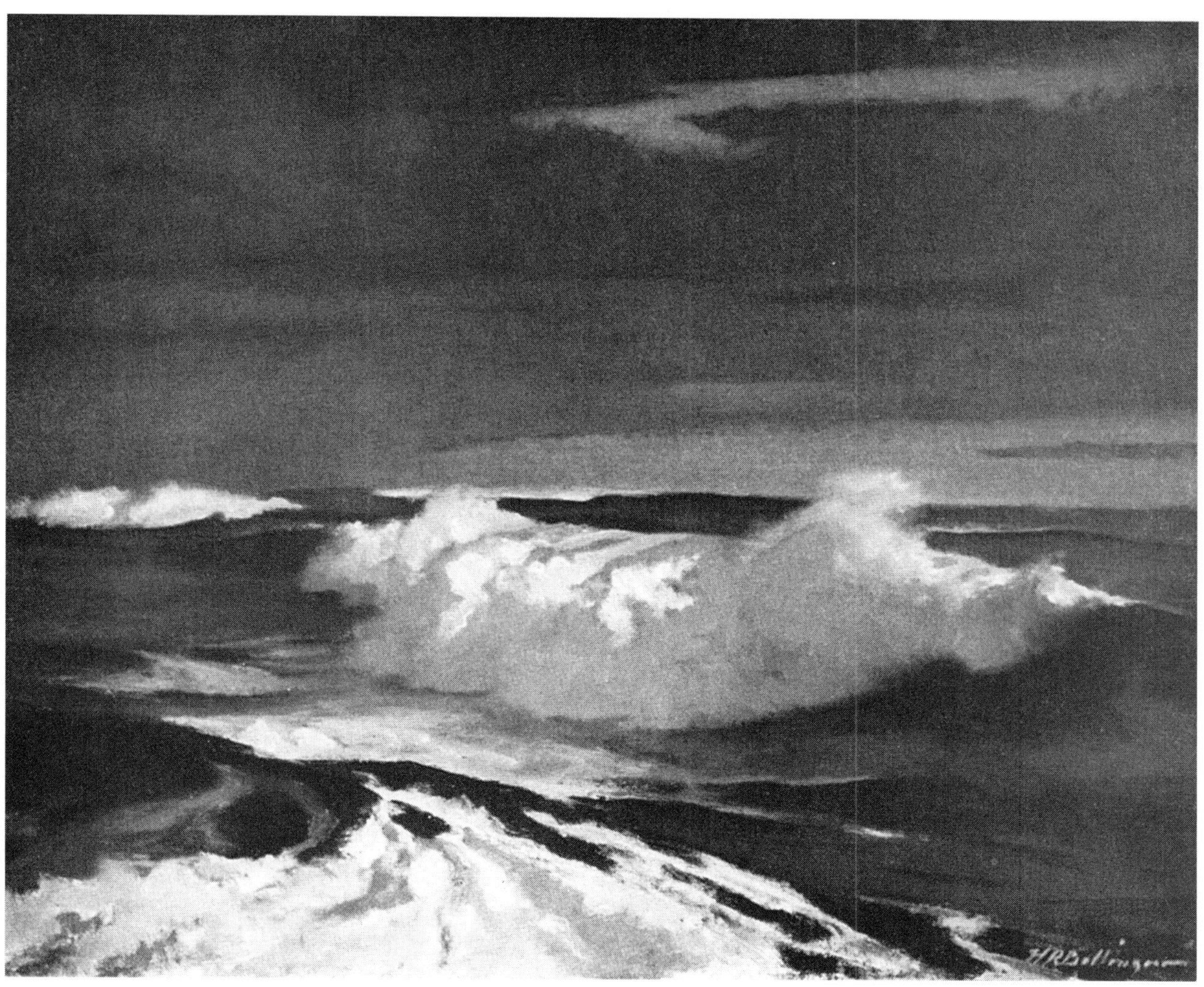

*A simple dark and light pattern broadly painted, oil, 16 x 20*

(See color insert between pages 42 and 43.)

# 11: The Sea by Night and Day

I think that if you want to do a great many pictures of the sea, it is fun to paint it in various moods and in different light effects. This will help you get away from the usual seascape with its nice blue sky, green sea, and brownish rocks. One of the most fascinating effects in nature is moonlight. It is lovely anywhere, but over the sea it is particularly striking and is well worth trying to paint. Unfortunately, it has to be done from memory, but I think that if you start your picture as soon as possible after studying the moonlit sea, you will still have a rather complete impression of the scene.

Moonlight is cooler and much weaker than sunlight. As a result, the shadows are darker and simpler. Light spots have little detail in them and shadows practically none. You can see big divisions of light and shade, but that is about all. In painting a night scene you can keep quite a little warmth in the darks and a little in the moonlight itself. I try to avoid having the whole picture too cold a green or blue because there is a lot more warmth in the moonlight than you think.

I generally paint the sea, sky, foam, and rocks in varying shades of warm blue green that sometimes verge on purple. The strong darks in the foreground should be warmer than those farther back in the picture. And often I paint a moonlight marine with a back light hitting on the sea and silhouetting waves and rocks against the light. If you don't show the moon itself, you can suggest its location just above the top of the picture either by showing its light on clouds near the top of the picture or by painting a blur of light in the sky itself, extending down into the picture.

When you don't show the source of light, you can go as high in key as you wish on the light hitting into your picture. If you show the moon itself, the rest of the picture has to be toned down so much in order to make the moon look bright that you have very little sparkle left.

With the exception of the brilliant path of moonlight down through the middle of your picture, most edges will be soft or entirely lost. It will be difficult to see the horizon line except in the path of moonlight, and even the edges of the light foam in shadow will be soft and almost lost.

All the lost edges add greatly to the effect of mystery which is part of the charm of a moonlight scene (Fig. 59).

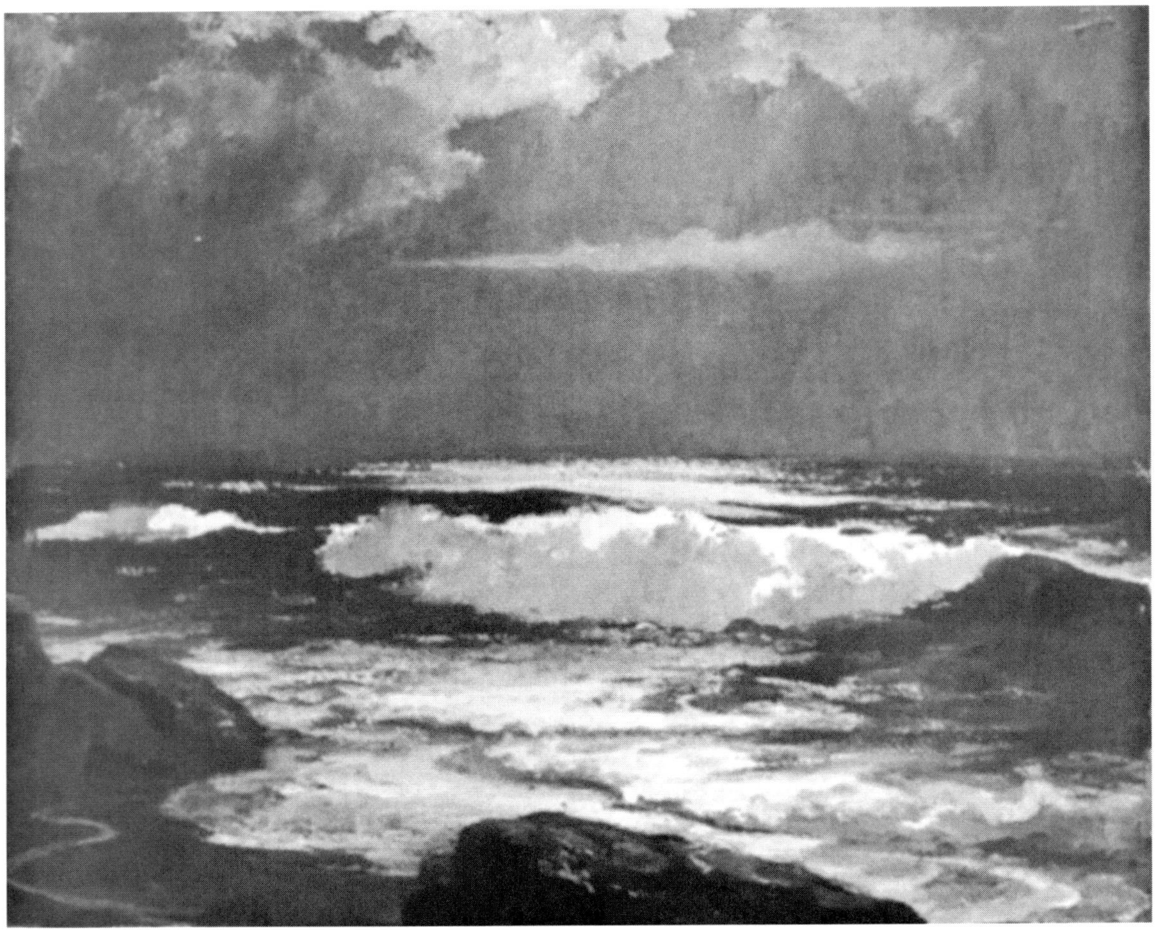

Fig. 59. *Moonlight on surf along the shore*

Though sunset and twilight scenes are sometimes considered rather sweet and sentimental, I like to paint them and I believe that everyone enjoys looking at them.

In a sunset picture the sky is the real center of interest and the sea is only important in that it reflects some of the brilliance of the sky. You usually look into the sunset to get the most brilliant color so, of course, you have a back lighting with much of the sea and rocks in silhouette, though you may have a rim of light around the tops of waves and rocks. There will be considerable light reflected from the sky on the horizontal planes of the sea and rocks, and I think it is better not to show the sun itself but to have it behind a cloud and have the light reflecting on the clouds and sea.

Here again most of your painting has to be done from memory, as a sunset changes every moment and practically never lasts long enough for you to make more than a few hasty notes of the color and general shapes of your most important clouds. As the sea and foreground are usually part of your dark pattern, you can have a fine, dark base in contrast to the light areas of the sky.

Sometimes it is advisable to tone down a little of the color that you see in the sky in order to keep the picture from looking too gaudy and too much like calendar art. I'm sure we have all seen effects in nature that would be unbelievable if reproduced in a picture.

Twilight for me has a great appeal. I like the simple pattern of light and shade that you see just before dark. The sky and light portions of the sea are parts of your light pattern; rocks and dark, uncrested waves make a fine, simple pattern of darks. The whole scene can almost be painted in

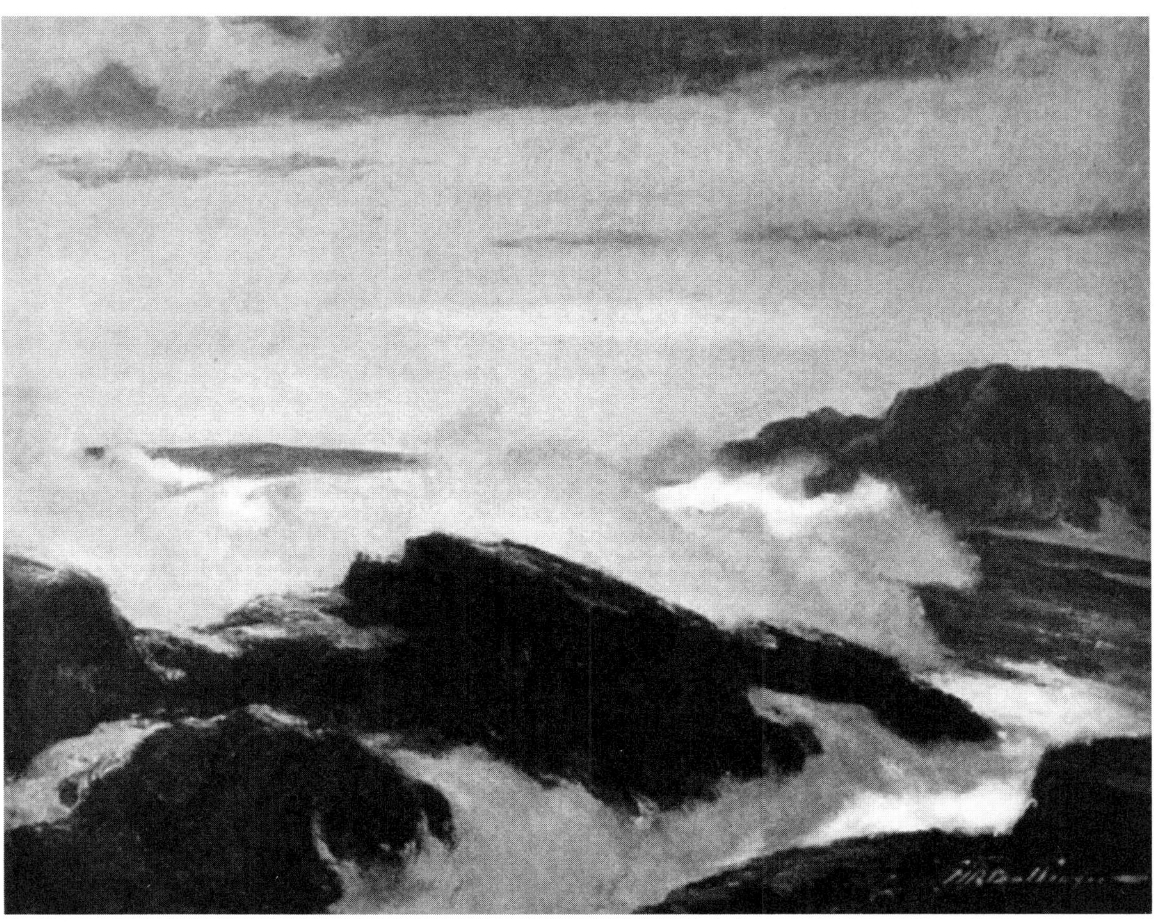

Fig. 60. *Twilight—rocky headland*

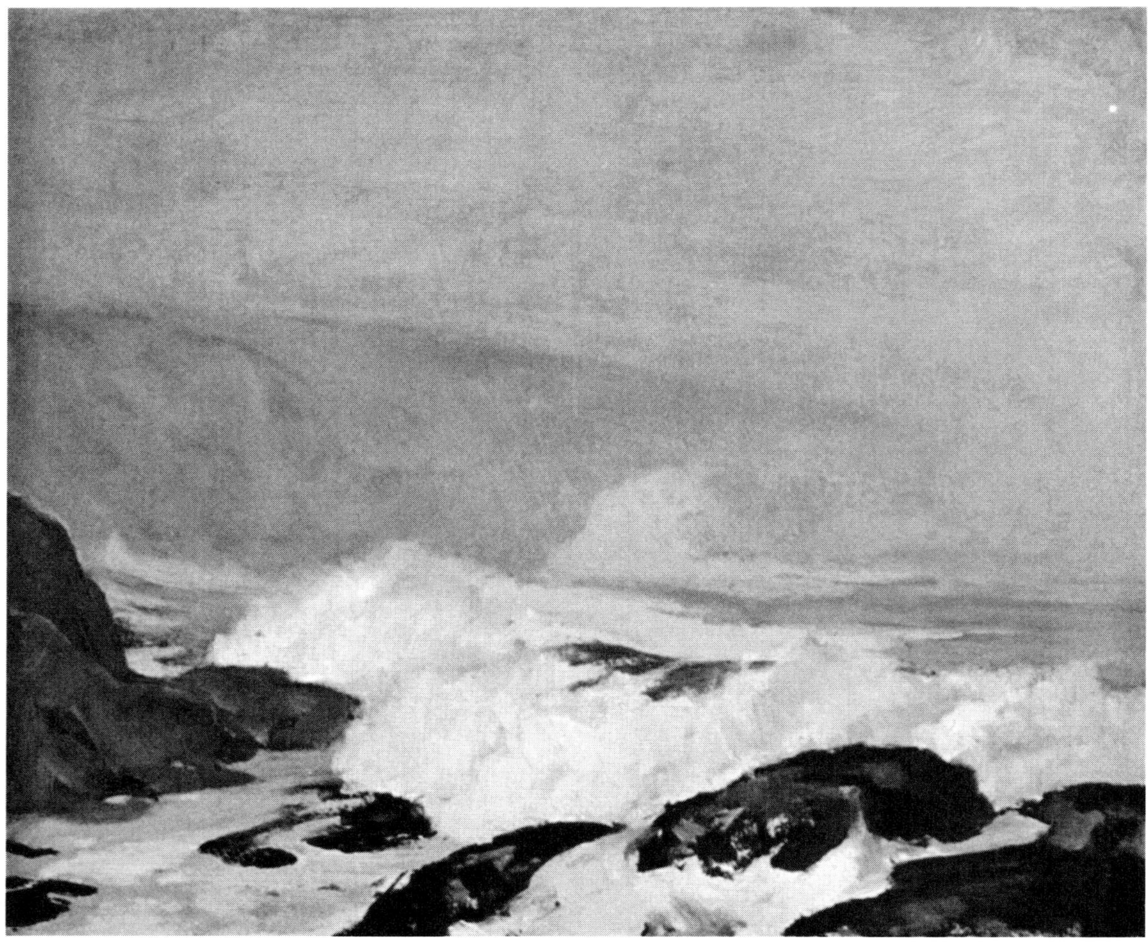

Fig. 61. *Fog scene*

two tones of light and shade (Fig. 60). There is very little detail anywhere —either in the lights or the darks—and you can paint the whole scene with few colors.

The more colors you use, the more complicated your picture becomes. Some of the strongest and most effective pictures are those painted with an analogous color scheme; that is, just a few related warm colors or a cool scheme with cool colors. In either, a single touch of a complementary color will make an effective accent.

I will sometimes paint an entire picture in tones of warm pink or tan with just a touch of cool green or blue as an accent. The result looks surprisingly convincing and gives the effect of full color.

Scenes of fog or rain are fascinating to paint and frequently well worth the effort involved.

A fog scene has mystery and can be very effective especially if you suggest a little warm light breaking through in the foreground or center of interest. The horizon line is practically lost and rocks and waves a little distance back in the picture can be painted in very pale gray spots of light and shade. It is sometimes a good idea to have a headland or some offshore rocks show dimly through the fog, thus emphasizing the misty quality of the atmosphere (Fig. 61).

You can paint a fog scene with a diffused down light on everything in the picture, thus emphasizing the light horizontal top planes in contrast to the darker upright planes of your foam, waves, and rocks. The whole picture can be painted in different tones of warm gray with a few warm touches of local color in the rocks or sand and a few greens or blues in the water.

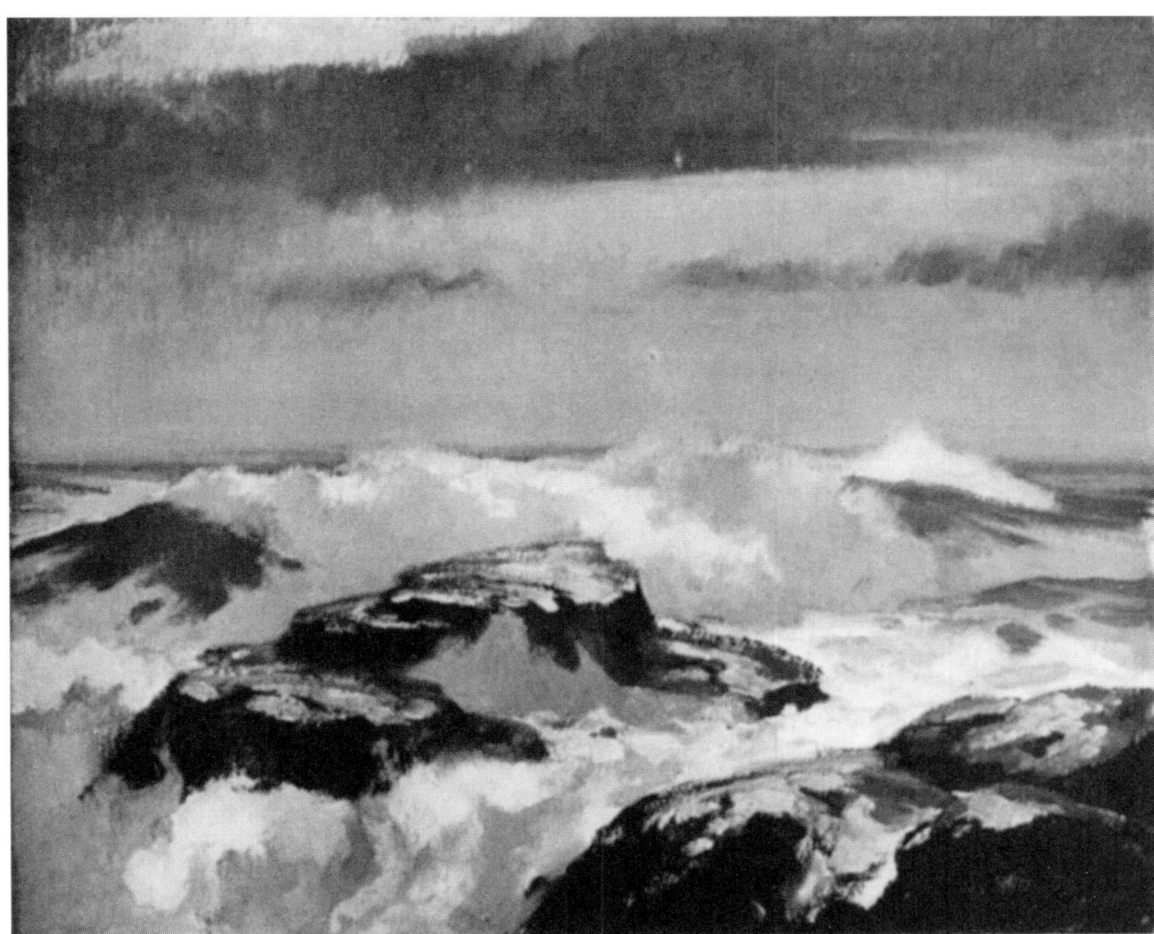

Fig. 62. *Rainy day—dark clouds*

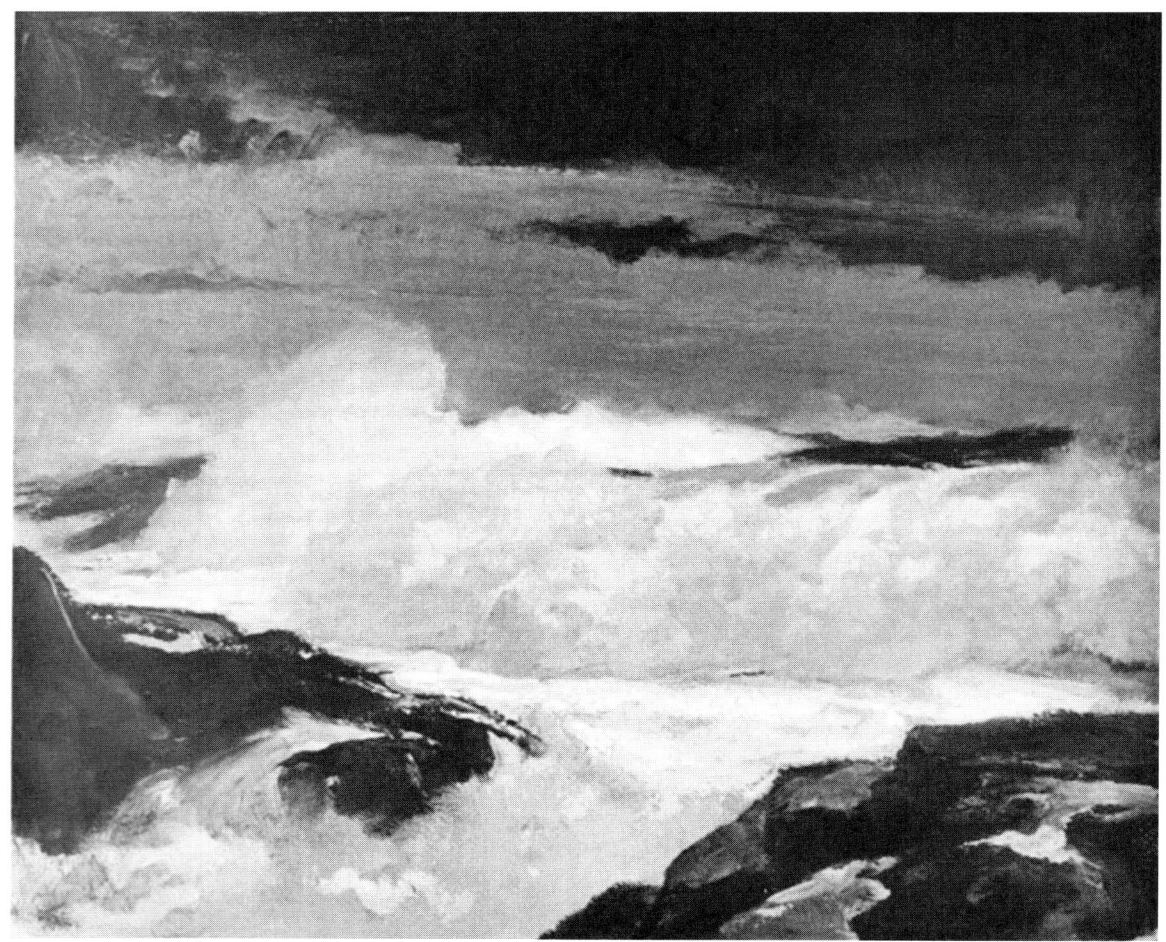

Fig. 63. *Storm scene*

A rainy-day picture is much like a fog scene except that you have a little less action in the surf and a wetter look to the rocks or sand. Also there are usually a few darker clouds in the sky (Fig. 62).

When you paint a storm scene you can really have fun. Here again, it is generally impossible to make more than a few brief notes out by the sea. About all one can do is to study the storm scene and then try to paint it later in the studio under more peaceful conditions (Fig. 63). It is sometimes possible to make pencil notes on the color or general action of the waves to help you remember the scene. When I was younger and more courageous I used to try to paint outside in a gale. But now I find that discretion is the better part of valor and I am perfectly contented to try to paint the wilder aspects of nature in more comfortable surroundings.

When painting any seascape the nearer you can get to the level of the water the better, particularly in a storm scene when the oncoming waves loom up higher and higher against the sky as they draw near. A great wave rushing towards one is menacing and terrifying and adds greatly to the drama that we are trying to portray.

You can dramatize the sky by painting storm clouds even darker than you see them; also try suggesting wind and movement by the twisted shapes of the cloud formations.

In a real storm there is generally a solid mass of foam close in to the rocks with no spots of clear water visible. After each wave comes crashing up over them, the rocks are dark and wet from the pouring sheets of water. When a wave strikes a rock there is generally a great burst of foam and spray. It would be well to save this violent action for your center of interest. There is generally a heavy backwash as the water pours off the rocks and is pulled back out to sea to meet the next oncoming wave.

I think that a storm scene usually looks more effective in an exhibition than on a living room wall; a tranquil scene is generally more restful to live with.

While studio painting is all right, there is still nothing that can take the place of working directly from nature. You will get inspiration from the sea itself that can never be achieved indoors. Most of my own pictures are painted at the shore though sometimes I have to put on the finishing touches in the studio.

I find that the paintings done on location are always a source of valuable information to me when I am away from the sea. While a great many fine marine pictures are painted in the studio, either from memory or from sketches made at the shore, you will find that it has been the information gained from making dozens of careful sketches at the seashore that made possible the more imaginative studio pictures.

# 12: Some Practice Subjects

In the next few pages there are ten typical seascapes that you can use as material for some practice pictures of your own. In five of them, I have planned the direction of light and the black-and-white pattern. As the

COMPOSITION NO. 1

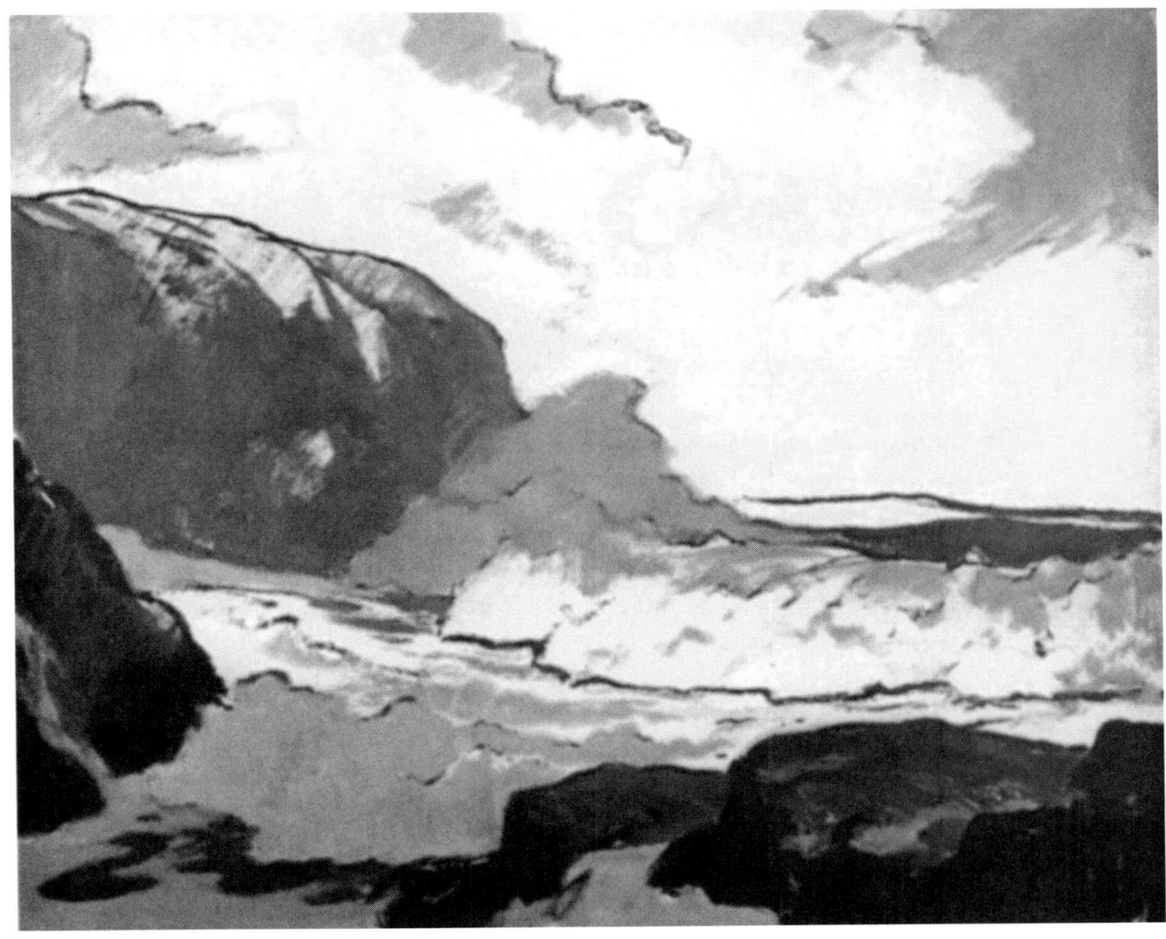

values have been indicated, it will only be necessary for you to figure out the color scheme you wish to use.

The remaining five compositions are only outline drawings, and will be a little bit more of a problem to develop into pictures. In addition to choosing the color schemes, you will also have to decide on the direction of light, the mood, and the black-and-white pattern of each picture.

I would suggest that you make a number of small composition notes before starting the large picture. These will help you find the best possible arrangement of your black-and-white pattern.

A possible treatment for each picture has been suggested, but I think you will enjoy trying out your own ideas on some of the sketches. You will find that you can produce a variety of pictures by changing the color scheme, direction of light, or the mood. Sometimes you can make an attractive moonlight or fog scene from a most ordinary looking seascape.

1. In the first composition of surf breaking on a rocky shore, you might use a warm light on your cresting wave in the center of the picture; perhaps some clouds in the sky. You could use a little of the color scheme of the step-by-step demonstration picture of surf cresting on a rocky shore, though I would like to see you use a little more blue in the sky, in the foreground rocks, and in the surf in shadow. The floating foam in sunlight could be quite warm and a little darker in value than the foam of the big, cresting wave. You might use some pale blue green in the sheet of water at the top of your big wave, and repeat it somewhere else in the picture where you have clear water in sunlight.

In painting the sky, you could have some big, swelling clouds building up in the center, with patches of blue sky where I have used the gray tone. Remember that with a front or side light, your clouds are lighter and more brilliant overhead, but are grayer and cooler as they near the horizon. Because of this, you will have to tone down the clouds as you work toward the middle of the picture. The distant wave in the center of the picture at the horizon is supposed to be in sunlight; it could be warmer and lighter than the rest of the sea around it.

2. In Composition No. 2, you might have a down light on the picture. The sky could be blue with more warmth in it as you near the source of light at the top of the picture. The clouds could have a warm light on their top planes, with cooler shadows underneath. Near the horizon line, you would have the usual pearly gray. The distant ocean would be a blue green. The headland would be a medium dark bluish purple. The rocks where the big burst of foam occurs on the upper right would be a little darker and warmer than the more distant cliffs on the left.

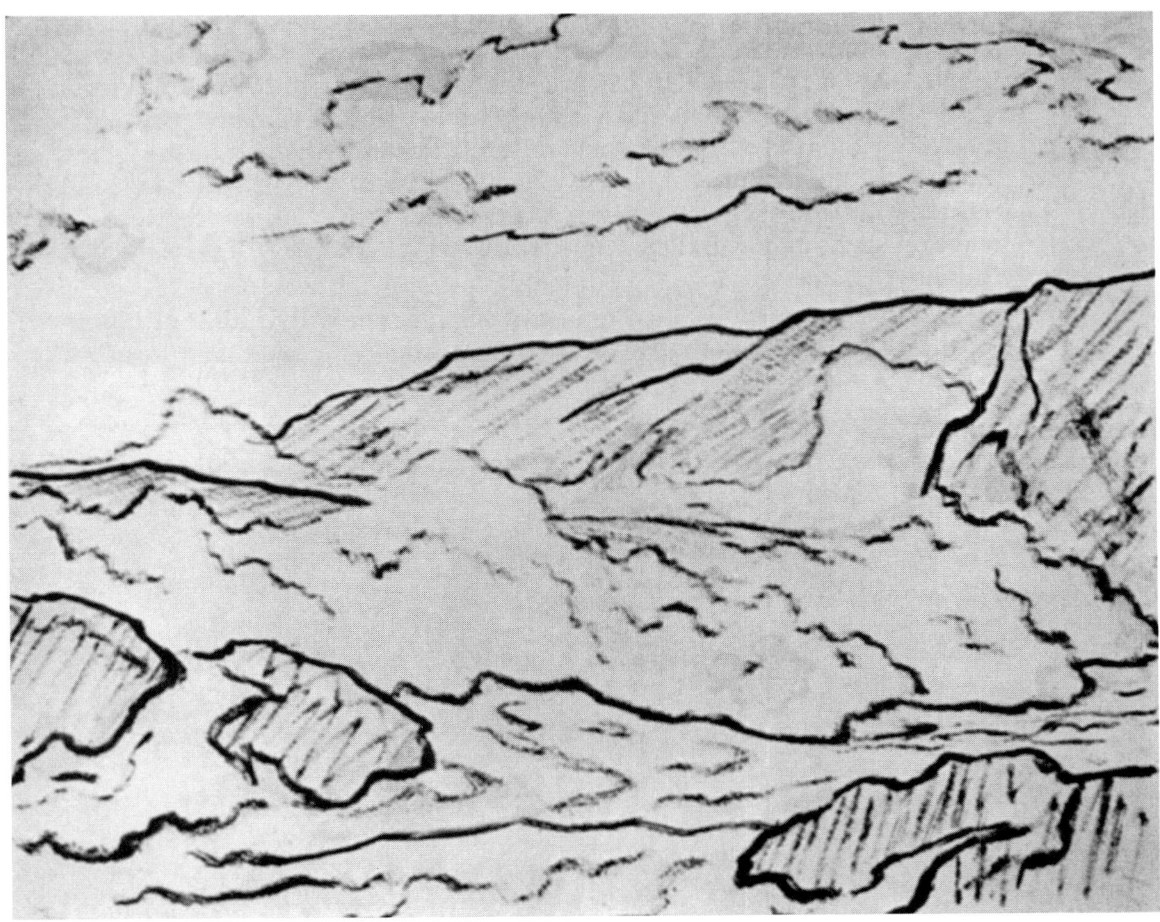

COMPOSITION NO. 2

The foreground rocks would be very dark with quite a good deal of warmth in them, especially in the shadows. However, a few of their top planes could reflect some of the cool color of the sky.

The foam of the cresting wave would have a little pale warmth in it, but there would be some delicate blue or lavender in the more upright portions of the foam. The floating foam in the foreground would be extremely light, with quite a little warmth. You could use just a touch of cadmium orange mixed with white. The water pouring over the rocks in the foreground would be quite light on the horizontal planes, then a little bluer or grayer as it pours down off the rocks. The big burst of foam on the right wouldn't be quite as warm or light as the big wave in the foreground.

3. In Composition No. 3—surf on a sandy beach—you have the light coming from the right, bathing the whole picture with warmth. I would suggest having pinkish clouds, some pink mixed through the blue sky, and a good deal of warmth in the cresting waves and the floating foam. The masses of foam that stand up above the surface of the water always catch more light and should be painted much higher in key than the horizontal portions of foam lying flat on the surface of the water. Be sure that the front edge of each of the flattened out waves is considerably lighter than the trailing foam behind it.

The distant sea could be a rich, dark blue green with a little of the gray of the sky reflecting in it. The clear water of the large cresting waves could

COMPOSITION NO. 3

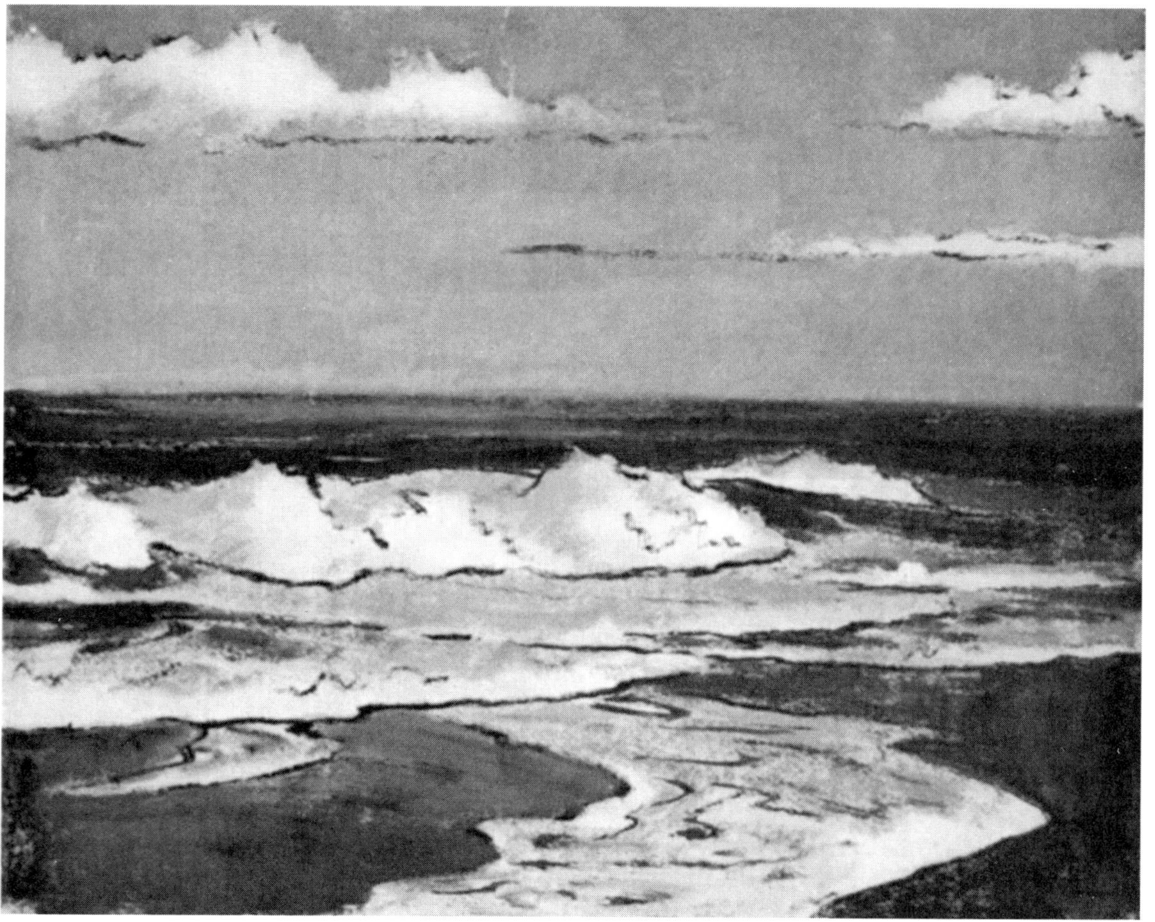

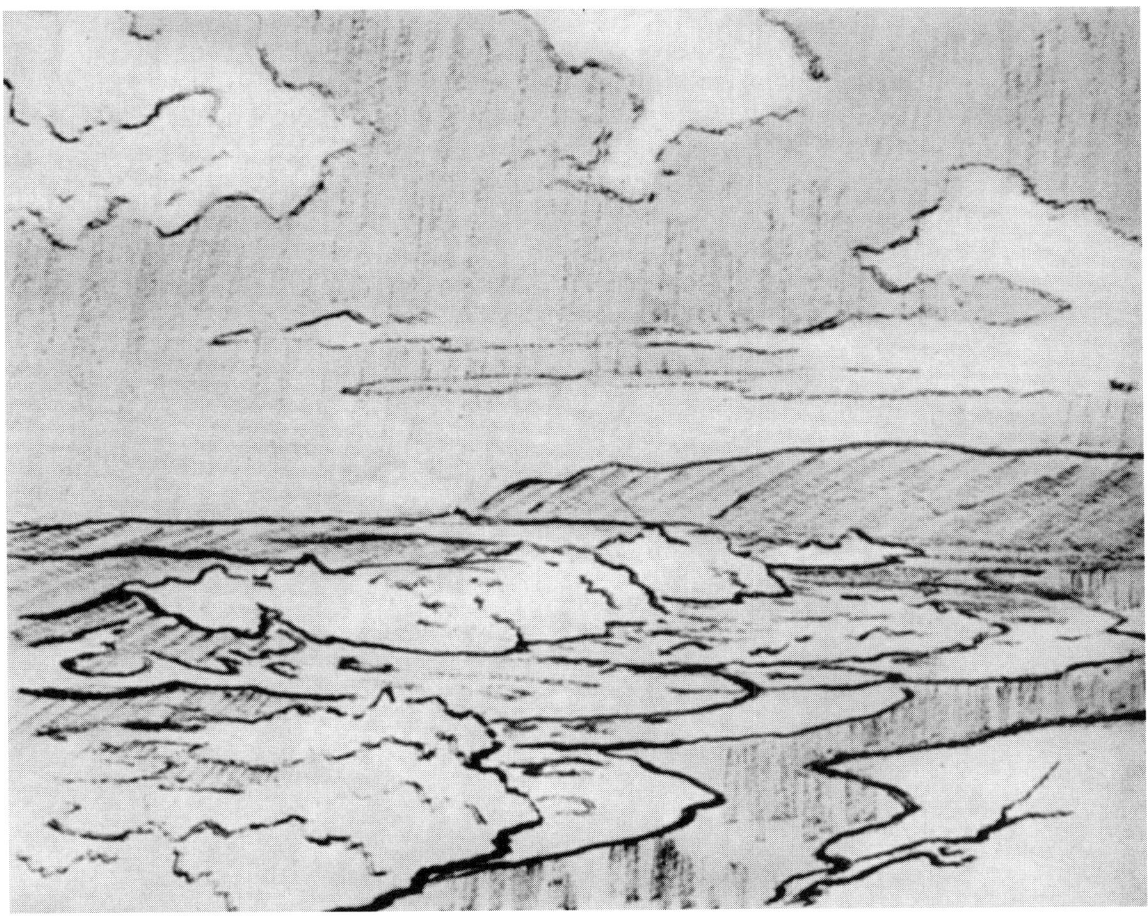

COMPOSITION NO. 4

be a little more green than blue, and the clear water between the masses of floating foam a lighter shade of green; while the color of the shallow water close to shore would be more the color of the sandy beach beneath it. The wet sand next to the water is more of a brownish purple, while the dry sand farther up the beach is a grayish tan and lighter in value. The trailing, floating foam generally picks up a little lavender from the sky.

4. In Composition No. 4—the second sandy beach scene—why not try a back lighting on the order of the step-by-step painting of surf breaking on a sandy beach. The headland could be a bluish purple and could be just a simple, dark tone with no detail in it. The front planes of all the cresting waves would be in shadow, with your strong lights on the top edges of the masses of foam, and on your flat horizontal planes.

The strip of wet sand next to the water can be painted quite dark, with a little light reflecting in it here and there to show that it is wet. Be sure to paint the reflections with up and down strokes. The sand in the right-hand corner could be a light grayish tan. The piece of driftwood could be a dark gray, or could have a few reddish brown touches.

The sun could be just above the top of the picture, a little to the right of center. Be sure to keep the sky and the center cloud in a blur of light. As you get farther away from the source of light, down nearer the horizon, you can paint the sky in a lower key. However, you would get a little warm light on the top edges of the lower clouds.

You can use a few touches of pale jade green in the dark sheet of water

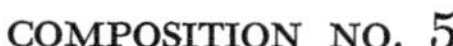

COMPOSITION NO. 5

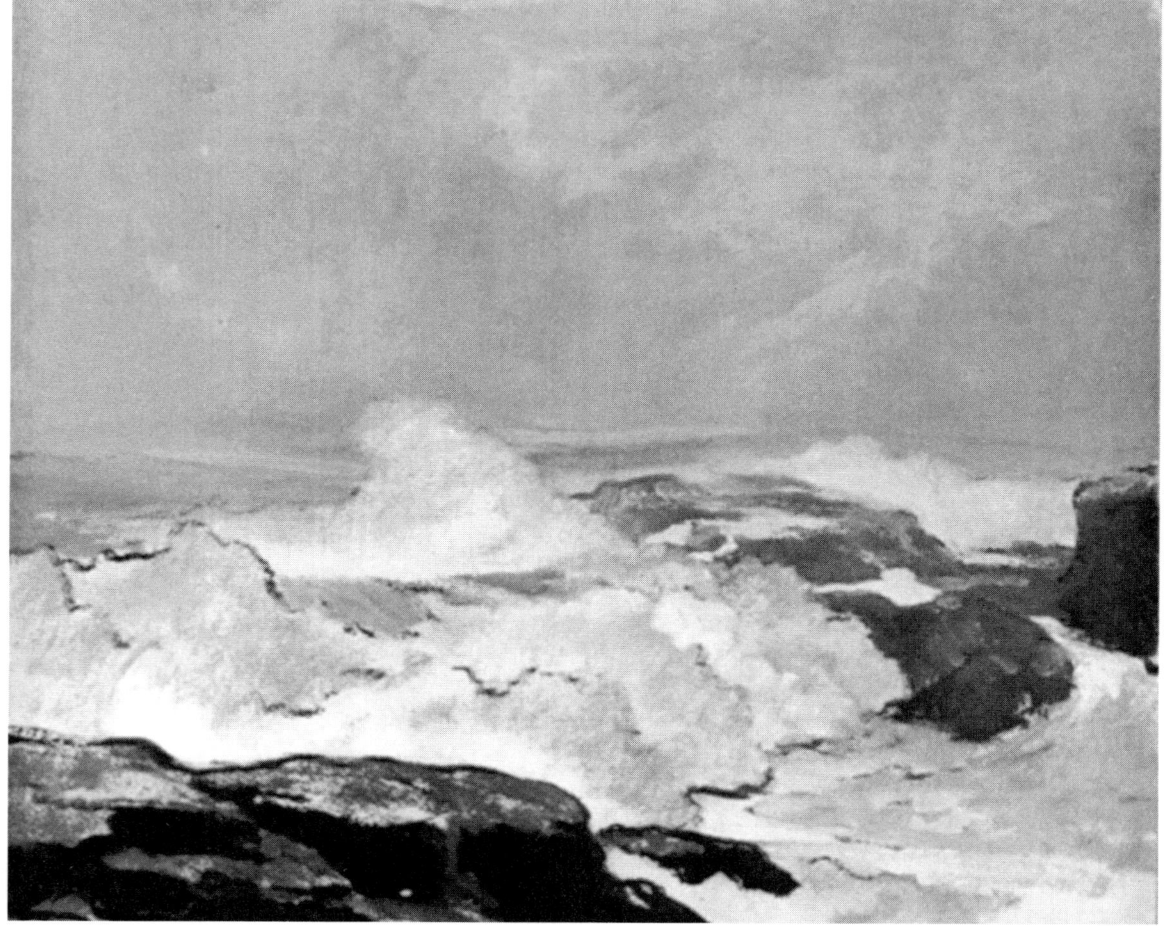

at the top of your important wave. Sometimes with a back lighting you will find that the light shining through the clear water at the top of a wave will produce a strong light green color. This will only be visible at the extreme top of the wave, where there is little thickness from front to back and the light can shine through the green sea water quite easily. Lower down on the wave, the back light can't penetrate and you see only the dark blue green of the wave itself.

5. Composition No. 5, is a fog scene and should be painted in tones of grayish tan. The horizon line should be lost and the masses of white foam in the distance should be just flat patterns of light, with hardly any detail and with very soft edges. You have a diffused down light on everything in the picture, with all of your upright planes a little darker than the horizontal ones.

If you paint your foreground rocks just as strong and dark as possible, you will be able to produce the effect of fog by painting everything in the background much grayer and softer.

The nearby surf and floating foam can be painted with plenty of detail to contrast with the half-lost shapes of the more distant waves. In the sky you can suggest fog by having the sky a little lighter at the top of the picture and having a few soft swirls of a slightly darker tone here and there in the sky.

6. Composition No. 6, is an open sea picture which I think might be painted with the light coming from the right side. The upper clouds could be a warm dark note; under that, you could have a band of blue sky getting a little warmer as you approach the source of light on the right. The lower bank of clouds would be a pink gray, though a little light might hit on the upper part of this cloud bank. The distant water would be a blue gray while the clear water in the foreground would be a rich, dark blue green. The swell in the lower left-hand corner should be the darkest spot in the picture and could be a very dark blue green. The clear water just above it with the floating foam could have a little more sky color in it and would not be quite as dark. The floating foam would be a warm cream color with perhaps a small amount of lavender added to it. It should be painted with your brush strokes following in the direction of the movement of the foam. The big cresting waves should be painted with up and down strokes to suggest weight and thickness. I would keep the foam of the breaking wave in the lower right-hand corner toned down with only a little light along the top of it. It isn't a good idea to have too much contrast in a corner of the picture, as it takes away from your center of interest. The cresting wave in the center would be quite light, with a little cool shadow on the left side of the foam masses. You will see how to model the foam of your big cresting

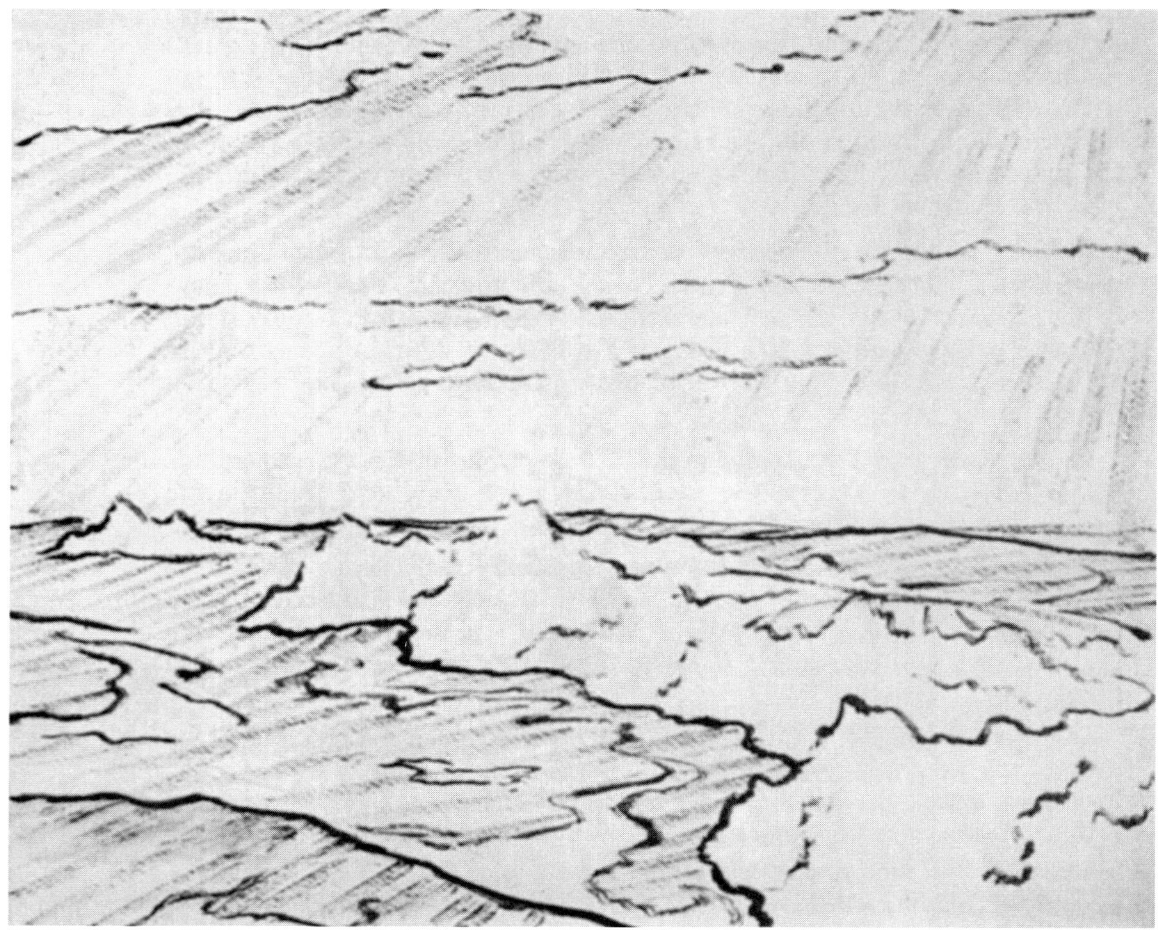

COMPOSITION NO. 6

wave by studying the color reproduction of the open sea demonstration picture.

7. Composition No. 7 is a moonlight scene with surf breaking on the rocky Monhegan coast. It should be painted in rich tones of rather grayed-up blue green with a good bit of warmth through the whole picture. Contrary to the usual belief, a moonlight scene isn't just a picture painted in cold greens and blues; it really has a lot of warmth in it, particularly in the shadows.

I would suggest that you have the moon just above the upper edge of the picture, but paint the upper clouds a pale grayish pink as though the moon were shining through them. This will enable you to have your brilliant moonlight reflected on the sea and on your foreground surf. You can have just a little warmth in the moonlight reflecting on the water.

The sea can be a shade darker than the sky, but be sure to have very soft edges at the horizon line. These soft edges will make the path of light reflecting on the water all the more brilliant and sparkling. The front planes of the cresting wave should be a grayed-up blue green with very little detail or modeling in them. In moonlight there is very little detail to be seen in the shadows. You can get them in the right shape and general value, but keep them just simple flat spots of tone.

I would try to lose as many edges as possible throughout the picture, only accenting the spots where the moonlight really hits. The foreground

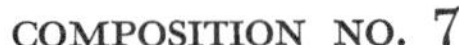

COMPOSITION NO. 7

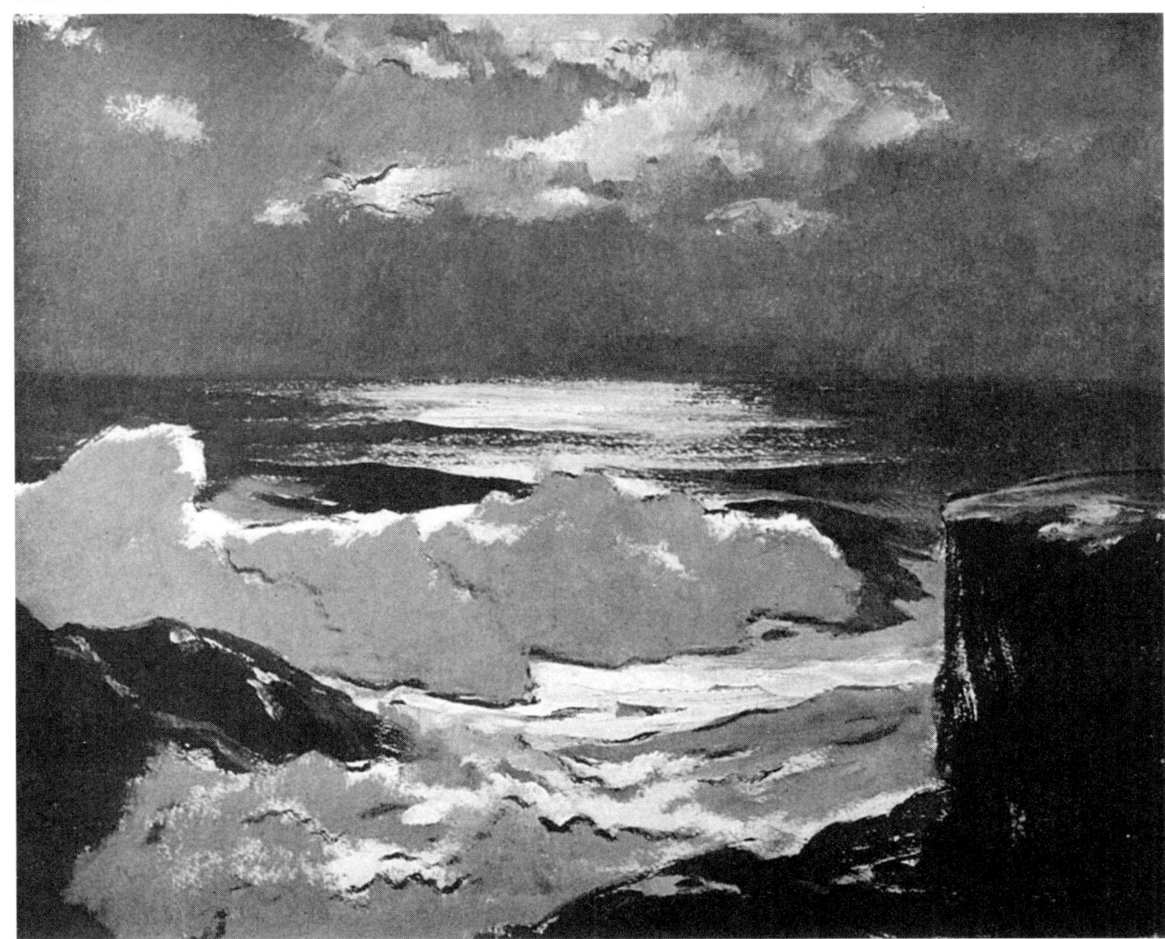

rocks in shadow should be a dark brownish blue, with a suggestion of warmth in the top planes that catch the light.

The great charm of a moonlight scene is in its illusive air of mystery, so be sure that you don't whittle everything out in too much detail.

8. Composition No. 8, can be another moonlight scene if you enjoyed doing the last one. Again I would have the moon above the top of the picture, a little right of center, with light striking the tops of the different cloud masses.

You could have light on the ocean and the floating foreground foam, also on the top edges of the cresting wave. The light on the freshly breaking wave would not be as brilliant as the light on the horizontal masses of foam.

COMPOSITION NO. 8

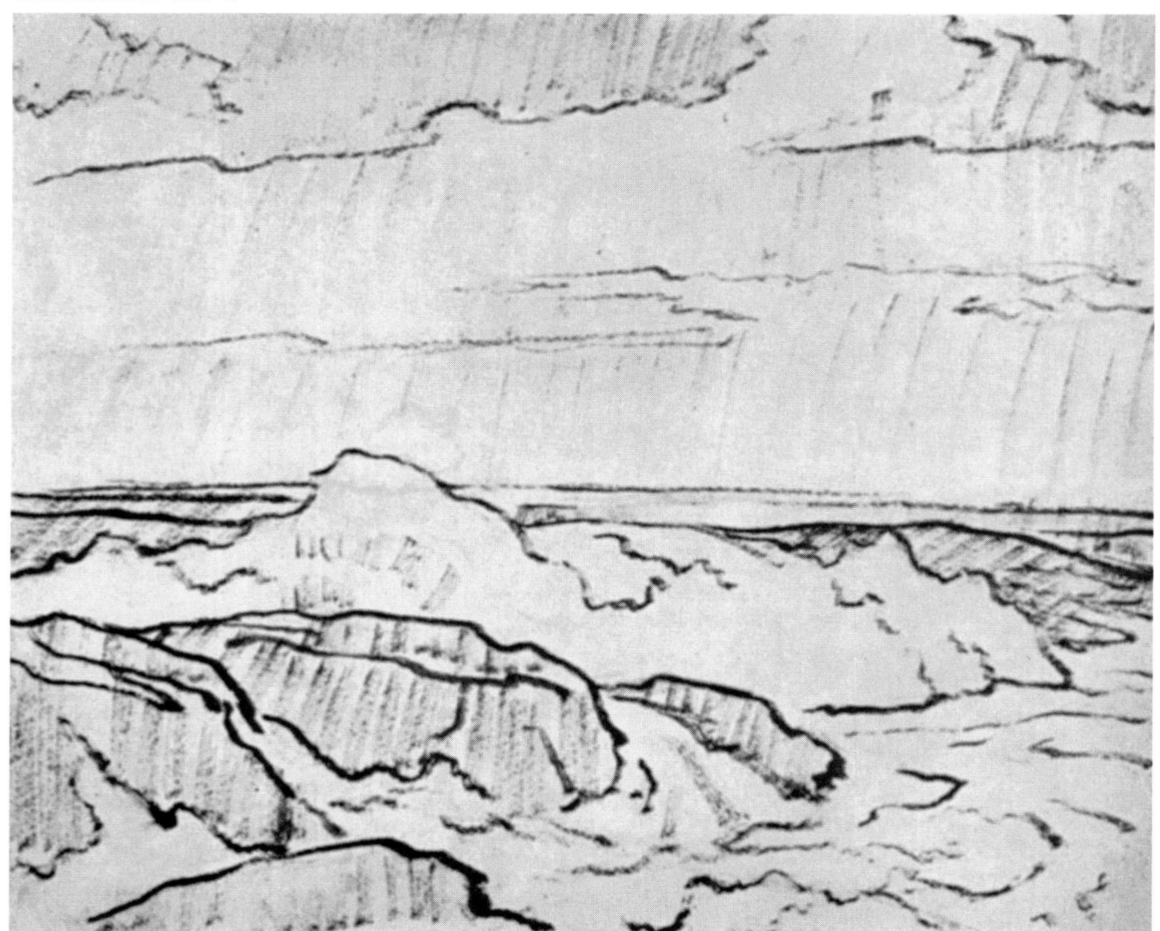

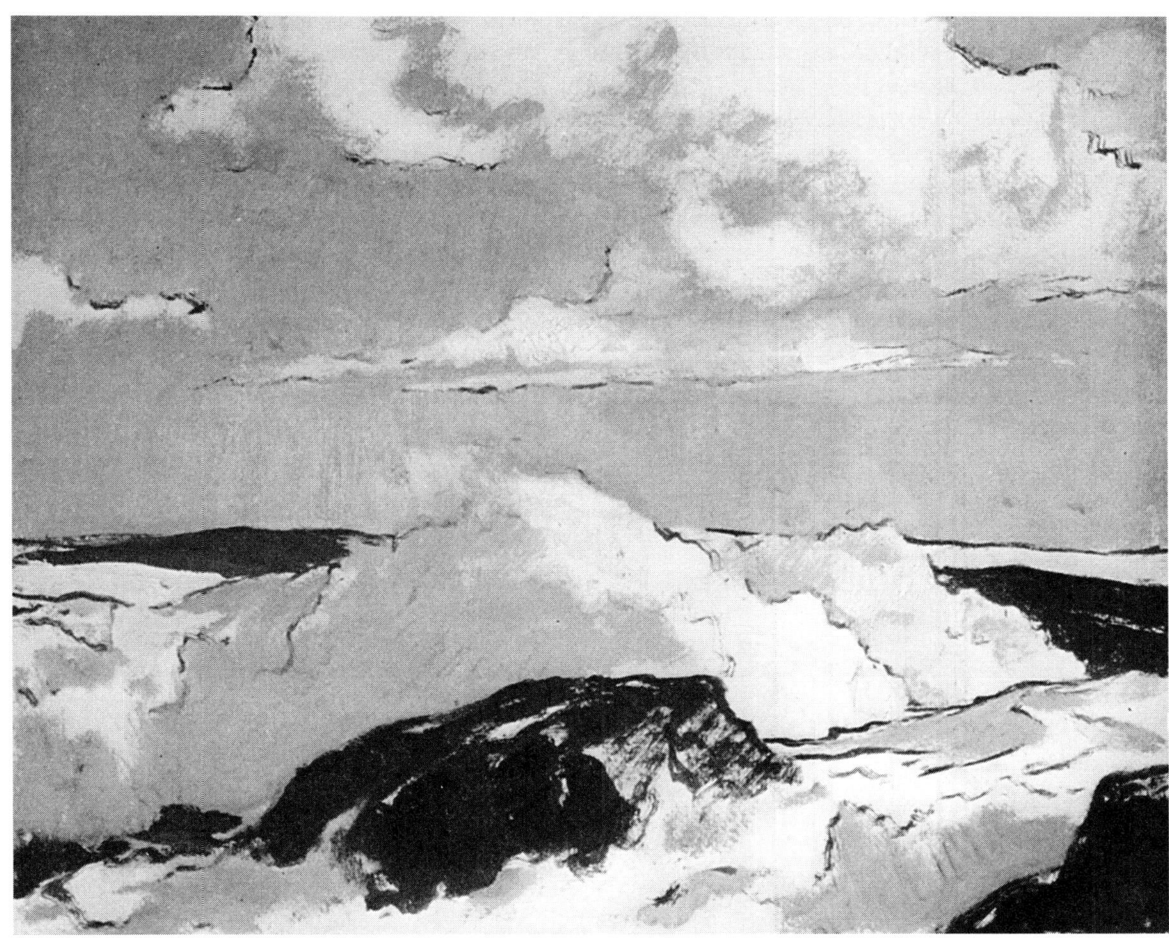

COMPOSITION NO. 9

Being broken up into so many planes, it doesn't reflect the light as well as the smoother, flatter horizontal surfaces.

The darks would be a dark purplish blue with just a little light on their top planes.

The foreground backwash would be quite light except at the lower left, where the water pouring off the rock would be in shadow.

You could model up the foreground floating foam to give it more weight and solidity. Remember that your upright planes are in shadow, while your strong light is on the more horizontal ones. Also, the front plane of your big, cresting wave would, of course, be all in shadow.

9. Composition No. 9 is a sunlit seascape with a nice blue sky and warm

pinkish or orange clouds. The sky should be more of an ultramarine blue at the top, fading into a light blue green halfway down to the horizon, then into a pink blue down near the sea. The dark wave at the left horizon and on the lower right would be a dark blue green.

The clear water in the lower right foreground could be a pale blue green, as though it were full of air bubbles, and you could use a grayed-up green for both the swell at the horizon on the right and the light wave halfway back on the left of the big burst of foam in the center. If a little of the sheet of water above the foam shows, that could be the same pale blue green.

The rocks would have some touches of warmth on the sunlit side; the shadow side would be a dark reddish brown with a little cool top light

COMPOSITION NO. 10

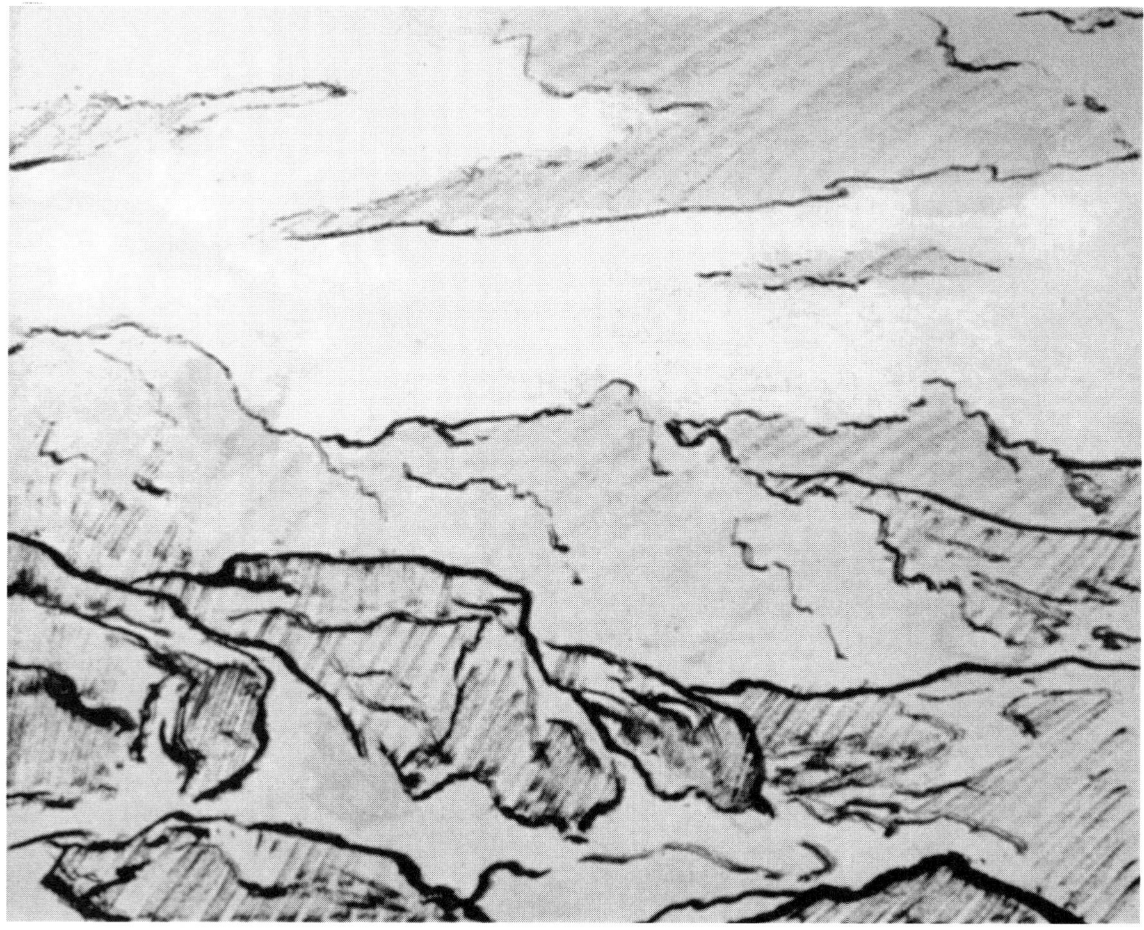

reflecting down onto the top planes in shadow. The foam of the big wave in sunlight would be quite warm; the floating foam would be even warmer and a little lower in key, except the portions of it that project above the level of the floating foam. These masses would, of course, be quite light.

10. For the last picture, with the great waves rolling into shore, I would like to have a twilight effect with a very light orange sky and everything else in the picture toned down. The clouds could be a warm, dark tan color and the foam of the cresting waves against the sky would be darker than the sky and would be a grayed-up blue green with a little warm top light reflected on the top edges of the foam masses. You would also get some of this reflected warmth in the horizontal portions of the foam pouring off the rocks and swirling back down to the sea. The clear water of the waves on the lower right-hand side would be a dark green with the floating foam a light tan color.

The foreground rocks would be a very dark purplish brown with a little warmth on their top planes. You could also have a little cool light from the sky overhead reflecting down into a few of the shadows here and there.

Be sure that your one really light spot in the picture is the lower part of the sky. Everything else must be toned down so that it will stay in its place and be part of a dark silhouette against the brilliance of the sky.

If you succeed with this type of picture, you will find that you have come a long way from the kind of seascape usually painted by the student artist.

# In Conclusion

I have always thought that marine painters were the luckiest of people, as they are painting the most magnificent subject in the world. However, you have to be really fond of the sea to go to all the trouble of learning to paint it. As you know, when you paint the ocean you are trying to portray motion as well as the mood and feeling of the scene before you. It isn't easy, but I think it is worth all the effort involved to be able to paint a really fine seascape—one that gives pleasure to yourself as well as your family and friends.

Many students admire the work of some individual artist so much that they are unduly influenced by his style or technique. They would like to paint exactly as he paints. I don't think this is a good idea. It isn't necessary to copy the style of any one artist, even though you may greatly admire his work.

It is a better plan to study the work of all of the marine painters, either at exhibitions or through reproductions of their paintings. You can get a great many ideas from their work that will help your own pictures. When you are influenced by the work of a number of artists, your painting won't resemble the work of any one painter, and will look entirely original.

When you are viewing the work of other artists try to look for the good points in their pictures. Never pay much attention to the faults. It is only the faults in your own pictures that matter. None of us are perfect painters, and I think it is easy to see faults in some of the finest works ever painted. Pictures are like friends. You like them in spite of their faults. So if there are a few spots in a picture that are less than perfect, think nothing of it as long as the general effect is good.

As regards your individual style and technique, you will find that if you paint in the manner and style that is most natural to you, you will begin to develop certain characteristic mannerisms that will distinguish your pictures from the work of everyone else.

Probably a number of you folks are like myself and can spend only a portion of the year at the seashore. In that case, when painting away from the water, you have to use as material either sketches you made at the shore or photographs of surf scenes. It is better to work from a black-and-white photograph than to copy someone else's painting. When you only copy another artist's picture you don't have to use your imagination, and consequently are not learning as much as when you are working out something original.

I have often been asked if I thought a person had to be born with a great talent in order to be a good painter. I don't think this is so, nor that you have to be a genius to paint a creditable picture.

Most people who have a desire to paint have some artistic talent, as a person rarely wants to do something for which he has no aptitude.

If you have ordinary common sense, good taste and the ability to work hard enough to learn the fundamentals of painting, you will be able to paint a picture. When you are painting you are training your eye and hand to see and record the scene before you.

There is nothing that can take the place of hard work. The only way to learn to paint is to constantly paint.

I have tried to include in this book all of the ideas on marine painting that I thought might be of help to the average painter. There is so much to learn about painting the surf and sea that I have always felt one lifetime wasn't long enough to learn all about it. I hope that this book will help the reader to avoid some of the long frustrating hours that I have had to spend in learning about marine painting, the hard way. Perhaps some day, if I am lucky, I may meet some of you nice people when we are both painting out on the shore, and you can tell me how you are getting along in your struggle with the "ole debble sea." Until then, may I wish you all the luck in the world with your own marine pictures, and many happy hours of painting by the sea.